Power Line Ampacity System

Theory, Modeling, and Applications

Anjan K. Deb, Ph.D., P.E.
Electrotech Consultant

CRC Press
Boca Raton London New York Washington, D.C.

Library of Congress Cataloging-in-Publication Data

Deb, Anjan K.
    Powerline ampacity system : theory, modeling, and applications / Anjan K. Deb.
       p. cm.
    Includes bibliographical references and index.
    ISBN 0-8493-1306-6
    1. Powerline ampacity—Mathematical models. 2. Electric cables—Evaluation. 3. Electric lines—Evaluation—Mathematical models. 4. Electric power systems—Load dispatching. 5. Electric currents—Measurement—Mathematics. 6. Amperes. I. Title.
    TK3307 .D35 2000
    621.319—dc21   00-036093
                                                                                                    CIP

This book contains information obtained from authentic and highly regarded sources. Reprinted material is quoted with permission, and sources are indicated. A wide variety of references are listed. Reasonable efforts have been made to publish reliable data and information, but the author and the publisher cannot assume responsibility for the validity of all materials or for the consequences of their use.

Neither this book nor any part may be reproduced or transmitted in any form or by any means, electronic or mechanical, including photocopying, microfilming, and recording, or by any information storage or retrieval system, without prior permission in writing from the publisher.

The consent of CRC Press LLC does not extend to copying for general distribution, for promotion, for creating new works, or for resale. Specific permission must be obtained in writing from CRC Press LLC for such copying.

Direct all inquiries to CRC Press LLC, 2000 N.W. Corporate Blvd., Boca Raton, Florida 33431.

Trademark Notice: Product or corporate names may be trademarks or registered trademarks, and are used only for identification and explanation, without intent to infringe.

© 2000 by CRC Press LLC

No claim to original U.S. Government works
International Standard Book Number 0-8493-1306-6
Library of Congress Card Number 00-036093
Printed in the United States of America 1 2 3 4 5 6 7 8 9 0
Printed on acid-free paper

Dedication

*This book is dedicated to my wife Meeta and
my family and friends*

About the Author

Dr. Anjan K. Deb is a registered professional electrical engineer in the state of California, and is a principal in ELECTROTECH Consultant, a transmission line software and consulting company that he started in 1990. He has 20 years' experience in high-voltage power transmission lines, substation automation, and electric power systems. He has authored or co-authored more than 20 research publications in the area of transmission line conductor thermal ratings, and has been awarded a U.S. patent for the invention of the LINEAMPS™ program.

Dr. Deb works as a consultant for electric power companies in all regions of the world, offering seminars and custom software solutions for increasing transmission line capacity by dynamic thermal ratings. As stated in this book, the LINEAMPS software developed by the author is used in several countries.

After receiving a bachelor's degree in electrical engineering from MACT India, Dr. Deb began his transmission line engineering career at EMC India, where he worked on the research and development of conductors and line hardware. He went to Algiers to work for the National Electrical and Electronics Company, where he designed and manufactured high-voltage substations. While working in Algeria, he received a French government scholarship to study Electrotechnique at the Conservatoire National des Arts et Métiers, Paris, France, where he received the equivalent of a master's degree in electrical engineering. He then received training at the Electricité de France (EDF) Research Center at Paris. EDF is the national electric power supply company of France. At EDF, Dr. Deb performed theoretical and experimental research on the heating of conductors and transmission line ampacity.

Dr. Deb came to the U.S. and began working for Pacific Gas & Electric (PG&E), San Francisco, where he developed and successfully implemented a real-time line-rating system for PG&E. While working at PG&E, he joined a doctoral degree program at the Columbia Pacific University, and earned a Ph.D. after completing all courses and preparing a doctoral dissertation on the subject of transmission line ampacity.

In addition to solving transmission line electrical and mechanical problems, Dr. Deb is interested in adaptive forecasting, energy management and developing intelligent computer applications for power. He is presently working on projects related to intelligent software development by the application of artificial intelligence, expert systems, object-oriented modeling, fuzzy sets, and neural networks. He also maintains the LINEAMPS website for interaction with program users, and for reporting new developments. He can be reached by e-mail at akdeb@aol.com, and on the Web at http://www.lineamps.com.

Preface

It is my great pleasure to present a book on transmission line ampacity. While there are several books devoted to transmission line voltage, there are few books that focus on line currents, computer modeling of line ampacity with power system applications, and the environmental impact of high currents. A unique contribution of this book is the development of a complete theory for the calculation of transmission line ampacity suitable for steady state operation and dynamic and transient conditions. To bring this theory into practice I have developed an object-model of the line ampacity system and implemented a declarative style of programming by rules. The end product is a state-of-the-art, user-friendly windows program with a good graphical user interface that can be used easily in all geographic regions.

As we enter the 21st century we shall have to develop new methods to maximize the capacity of existing transmission and distribution facilities. The power system may have to be operated more closely to generation stability limits for better utilization of existing facilities. Adding new lines will become more difficult as public awareness of environmental protection and land use increases.

To keep pace with increasing electric energy usage in the next millennium, new lines will be required for more efficient electricity transmission and distribution. Hopefully, with the help of material presented in this book, the transmission line engineer will make better decisions regarding the choice of conductors, environmental impact, system operation, and cost optimization.

This book is primarily for practicing electric power company engineers and consultants who are responsible for the planning, operations, design, construction, and costing of overhead powerlines. It is also a useful source of reference for various government authorities, electricity regulators, and electric energy policy makers who want to get a firm grip on technical issues concerning the movement of electric energy from one location to another, environmental concerns, and up-to-date knowledge of existing and future transmission line technologies.

Academicians and students will find material covering theoretical concepts of conductor thermal modeling, the analysis of conductor ampacity, powerline EMF developed from Maxwell's equations and Ampere's law, power flow with variable line ratings, stability analysis, power electronic devices, and flexible AC transmission. These will complement the existing large number of excellent textbooks on electric power systems.

This book has been developed from more than 20 years of my experience in working with various electric power companies in Asia, Africa, Europe, and North America. I am particularly grateful to Electricité de France, Paris, for the various interactions with the members of the Departement Etudes et Recherche since 1978, where I initiated research on the heating of conductors. Pacific Gas & Electric Company, San Ramon, California, offered me an excellent environment for research and development when I worked as a consultant on transmission line dynamic thermal ratings.

Thanks are due to several users of the LINEAMPS program, including TransPower, New Zealand; Hydro Quebec, Canada; and Korea Electric Power Company for their valuable feedback and support which has enabled me to enhance the computer program. The kind technical support offered by Mr. Graham of Intellicorp, California, during the development of the LINEAMPS program is gratefully acknowledged.

Thanks are due to Dr. Peter Pick, Dean, and Dr. John Heldt, Mentor, of Columbia Pacific University, California, for their guidance while I prepared a doctoral dissertation, and for their continued encouragement to write this book. I thank Ms. Genevieve Gauthier, Research Engineer, Institute de Recherche Hydro Quebec, Canada, for going through the initial manuscript and kindly pointing out errors and omissions. Last, but not the least, I am grateful to Professor R. Bonnefille, University of Paris VI, and Professor J. F. Rialland for their lectures and teachings on Electrotechnique at the Conservatoire National des Arts et Métiers, Paris, France.

As in most modern electrical engineering books, SI (System International) units are used consistently throughout. Complex numbers are denoted by an upperscore, for example, a complex current $I \angle \theta = I \cdot e^{j\theta}$ is represented by $\bar{I}$, and a vector is denoted by an upper arrow like $\vec{H}$.*

The LINEAMPS computer program described in this book is a commercial software program available from:

ELECTROTECH Consultant
4221 Minstrell Lane
Fairfax, VA 22033, USA
(703) 322-8345

For additional details of the program and to obtain new information concerning recent developments on high currents and transmission line ampacity, readers may visit LINEAMPS on the Web at http://www.lineamps.com.

Anjan K. Deb, Ph.D.

---

* I have followed the same notation used by Gayle F. Miner in *Lines and Electromagnetic Fields for Engineers*, Oxford University Press, New York, 1996.

Contents

Chapter 1 Introduction 1
    1.1   Organization of Book and Chapter Description 1
    1.2   Introducing the Powerline Ampacity System 3
    1.3   Electric Power System Overview 3
            1.3.1   Transmission Grid 3
            1.3.2   Overhead Transmission Line 4
            1.3.3   High-Voltage Substation 8
            1.3.4   Energy Control Center 9
    1.4   Factors Affecting Transmission Capacity and
           Remedial Measures 11
    1.5   New Developments For Transmission Capacity Enhancement 12
    1.6   Dynamic Line Rating Cost–Benefit Analysis 12
    1.7   Chapter Summary 12

Chapter 2 Line Rating Methods 15
    2.1   Historical Backround 15
            2.1.1   Early Works on Conductor Thermal Rating 15
            2.1.2   IEEE and Cigré Standards 15
            2.1.3   Utility Practice 15
    2.2   Line Rating Methods 16
            2.2.1   Defining the Line Ampacity Problem 16
            2.2.2   Static and Dynamic Line Ratings 17
            2.2.3   Weather-Dependent Systems 18
            2.2.4   Online Temperature Monitoring System 19
            2.2.5   Online Tension Monitoring System 21
            2.2.6   Sag-Monitoring System 22
            2.2.7   Distributed Temperature Sensor System 23
            2.2.8   Object-Oriented Modeling and Expert Line
                      Rating System 25
    2.3   Chapter Summary 26

Chapter 3 Theory of Transmission Line Ampacity 27
    3.1   Introduction 27
    3.2   Conductor Thermal Modeling 28
            3.2.1   General Heat Equation 28
            3.2.2   Differential Equation of Conductor Temperature 29
            3.2.3   Steady-State Ampacity 29
            3.2.4   Dynamic Ampacity 36
            3.2.5   Transient Ampacity 44

	3.2.6	Radial Conductor Temperature	47
3.3		Chapter Summary	49
Appendix 3		AC Resistance of ACSR	51

Chapter 4 Experimental Verification of Transmission Line Ampacity 61
- 4.1 Introduction 61
- 4.2 Wind Tunnel Experiments 61
- 4.3 Experiment in Outdoor Test Span 63
- 4.4 Comparison of LINEAMPS with IEEE and Cigré 66
  - 4.4.1 Steady-State Ampacity 66
  - 4.4.2 Dynamic Ampacity 71
- 4.5 Measurement of Transmission Line Conductor Temperature ... 71
- 4.6 Chapter Summary 72

Chapter 5 Elevated Temperature Effects 73
- 5.1 Introduction 73
  - 5.1.1 Existing Programs 74
- 5.2 Transmission Line Sag and Tension — A Probabilistic Approach 74
  - 5.2.1 The Transmission Line Sag–Tension Problem 75
  - 5.2.2 Methodology 75
  - 5.2.3 Computer Programs 77
- 5.3 Change of State Equation 78
  - 5.3.1 Results 79
  - 5.3.2 Conductor Stress/Strain Relationship 80
- 5.4 Permanent Elongation of Conductor 80
  - 5.4.1 Geometric Settlement 81
  - 5.4.2 Metallurgical Creep 81
  - 5.4.3 Recursive Estimation of Permanent Elongation 82
- 5.5 Loss of Strength 83
  - 5.5.1 Percentile Method 83
  - 5.5.2 Recursive Estimation of Loss of Strength 84
- 5.6 Chapter Summary 84
- Appendix 5 Sag and Tension Calculations 87

Chapter 6 Transmission Line Electric and Magnetic Fields 93
- 6.1 Introduction 93
- 6.2 Transmission Line Magnetic Field 93
  - 6.2.1 The Magnetic Field of a Conductor 94
  - 6.2.2 The Magnetic Field of a Three-Phase Powerline 98
  - 6.2.3 The Magnetic Field of Different Transmission Line Geometry 100
  - 6.2.4 EMF Mitigation 102
- 6.3 Transmission Line Electric Field 108
- 6.4 Chapter Summary 113

Chapter 7 Weather Modeling for Forecasting Transmission Line Ampacity 115
   7.1   Introduction 115
   7.2   Fourier Series Model 116
   7.3   Real-Time Forecasting 123
   7.4   Artificial Neural Network Model 127
   7.5   Modeling by Fuzzy Sets 132
   7.6   Solar Radiation Model 137
   7.7   Chapter Summary 139

Chapter 8 Computer Modeling 143
   8.1   Introduction 143
        8.1.1   From Theory to Practice 143
        8.1.2   The LINEAMPS Expert System 143
   8.2   Object Model of Transmission Line Ampacity System 144
        8.2.1   LINEAMPS Object Model 144
        8.2.2   Transmission Line Object 145
        8.2.3   Weather Station Object 148
        8.2.4   Conductor Object 150
        8.2.5   Cartograph Object 152
   8.3   Expert System Design 154
        8.3.1   Goal-Oriented Programming 155
        8.3.2   Expert System Rules 157
   8.4   Program Description 159
        8.4.1   LINEAMPS Windows 159
        8.4.2   Modeling Transmission Line and Environment 159
        8.4.3   LINEAMPS Control Panel 159
   8.5   Chapter Summary 162

Chapter 9 New Methods of Increasing Transmission Capacity 163
   9.1   Introduction 163
   9.2   Advancement in Power Semiconductor Devices 163
   9.3   Flexible AC Transmission 168
   9.4   Chapter Summary 181

Chapter 10 Applications 183
   10.1   Introduction 183
   10.2   Economic Operation 183
        10.2.1 Formulation of the Optimization Problem 184
        10.2.2 Electricity Generation Cost Saving in Interconnected Transmission Network 186
   10.3   Stability 190
        10.3.1 Dynamic Stability 191
        10.3.2 Transient Stability 193
   10.4   Transmission Planning 195
   10.5   Long-Distance Transmission 198

	10.6	Protection	201
	10.7	Chapter Summary	204
Appendix 10.1		Transmission Line Equations	205

Chapter 11   Summary, Future Plans, and Conclusion 209
    11.1   Summary 209
    11.2   Main Contributions 212
    11.3   Suggestions for Future Work 219
    11.4   A Plan to Develop LINEAMPS for America 225
    11.5   Conclusion 226

Bibliography 229

Appendices A1–A8: Conductor Data 235

Appendix B: Wire Properties 243

Index 245

1 Introduction

1.1 ORGANIZATION OF BOOK AND CHAPTER DESCRIPTION

Chapter 1 gives a broad overview of the electric power system including transmission lines, substations, and energy control centers. Data for electricity production in the U.S. and the world are also given.

Chapter 2 presents the different methods of transmission line rating, including both on-line and off-line methods.

A complete theory of transmission line ampacity is presented in Chapter 3. A three-dimensional conductor thermal model is first developed, and then solutions are presented for steady-state, dynamic, and transient operating conditions.

Experimental work related to transmission line ampacity that was conducted in different research laboratories is described in Chapter 4. The conductor thermal models in the steady-state and dynamic and transient states are validated by comparing results with the IEEE standard and Cigré method. Results are also compared to laboratory experiments and measurements from actual transmission lines.

The effects of elevated temperature operation on transmission line conductors are presented in Chapter 5. Experimentally derived models of loss of tensile strength of conductors, as well as permanent elongation of conductors due to creep, are presented in this chapter. The method of calculation of the loss of strength and inelastic elongation of conductors by a recursive procedure that utilizes probability distribution of conductor temperature in service is described. A method of generating the probability distribution of conductor temperature in service from time series stochastic and deterministic models is given.

The theory of transmission line electric and magnetic fields is developed from Maxwell's equations in Chapter 6. When higher ampacity is allowed on the line, it increases the magnetic field radiated from the transmission line. The electric field from the transmission line does not change with line ampacity, but increases with conductor temperature due to lowering of the conductor to ground clearance by sag. Methods of reducing the level of EMF radiated from transmission lines by active and passive shielding are presented in this chapter. This aspect of transmission line ampacity is significant because there is little previous work carried out in this direction. Even though there is no evidence of environmental impact by EMF due to increased transmission line currents, measures are suggested to lower magnetic fields from transmission lines.

Environmental factors influence transmission capacity significantly. For this reason Chapter 7 is devoted entirely to weather modeling. The meteorological variables that are most important to powerline capacity are ambient temperature, wind speed, wind direction, and solar radiation. Statistical modeling of weather variables based on time series analysis, Fourier series analysis, and neural networks are presented with examples using real data collected from the National Weather Service. Models are developed for real-time prediction of weather variables from measurements as well as by weather pattern recognition. Analytical expressions for the calculation of solar radiation on a transmission line conductor are also presented to complete the chapter on weather modeling.

Chapter 8 describes computer modeling of the LINEAMPS expert system. The complete system of rating overhead powerlines is implemented in a computer program called LINEAMPS. This state-of-the-art software package is an expert system for the rating of powerlines. It was developed by object-oriented modeling and expert rules of powerline ampacity. The object of the program is to maximize the current-carrying capacity, or the ampacity, of existing and future overhead powerlines as functions of present and forecast weather conditions. Methods of object-oriented modeling of transmission lines, weather stations, and powerline conductors are described with examples from electric power companies in the different regions of the world. Expert system rules are developed to enable an intelligent powerline ampacity system to check user input and explain error messages like a true expert.

Chapter 9 discusses new methods of increasing line ampacity. The capacity of electric powerlines to transport electric energy from one point to another, that depends upon several factors, is discussed. The most important factors are transmission distance, voltage level, and generator stability. In many cases, adequate stability can be maintained by electrical control of generation systems as well as by fast control of active and reactive power supply to the system. When energy is transported over a long distance, there may be significant voltage drop that may be compensated for by controlling reactive power and/or boosting voltage levels by transformer action. Therefore, in most cases the transport capacity of overhead powerlines is limited only by the thermal rating of the powerline conductor. An overview of new technologies that are being developed to increase transmission capacity up to the thermal limit by overcoming the aforementioned limitations is presented in this chapter. These new technologies include the application of modern power electronics devices that are known as FACTS (Flexible AC Transmission System), Superconducting Magnetic Energy Storage (SMES), and distributed generation systems.

Chapter 10 presents applications of the new powerline ampacity system to clearly show its benefits. In a competitive power supply business environment, it is necessary to optimize the ampacity of overhead power transmission lines to enable the most economic power system operation on an hour-by-hour basis. Until recently, electric power companies* have assumed that the maximum capacity of a powerline is constant by assuming conservative weather conditions, so they followed a static line rating system. Now certain electric power companies** are adopting a system of line rating that is variable and dynamic depending upon actual weather conditions.

---

* *Regles de calcul electrique.* EDF/CERT *Directives Lignes Aeriennes* 1996.
  Ampacity of overhead line conductors. PG&E Engineering Standard.
** REE Spain (Soto et al., Cigre, 1998); KEPCO, South Korea (Wook et al., 1997)

Introduction

The thermal rating of a transmission line depends upon the maximum design temperature of the line, and the temperature of a conductor varies as a function of line current and meteorological conditions. Therefore, for the same value of maximum conductor temperature, higher line currents are possible if there are favorable meteorological conditions. In this chapter, a system of equations for the economic operation of diverse generation sources in an interconnected power system is developed that utilizes a dynamic line rating system. The economic benefits of a dynamic line rating system are demonstrated by giving an example of an interconnected transmission network having a diverse mix of electricity generation sources. The chapter concludes with a discussion of increased competition in the electric power supply industry in a power pool system of operations, and the important role of the powerline ampacity system presented in this book.

Chapter 11 gives a summary of main contributions made in this book, presents future plans and new transmission and distribution technologies, describes the role of Independent System Operators (ISO) and power-pool operations from the point of view of transmission line capacity. It provides a discussion on deregulation and how the line ampacity system facilitates greater competition in the electric supply business.

1.2 INTRODUCING THE POWERLINE AMPACITY SYSTEM

As the demand for electricity increases, there is a need to increase electricity generation, transmission, and distribution capacities to match demand. While the location and construction of a generation facility is relatively easy, it is becoming increasingly difficult to construct new lines. As a result, electric power authorities everywhere are searching for new ways to maximize the capacity of powerlines. One of the methods used to increase line capacity is dynamic thermal rating.

The object of this book is to develop a complete system of rating overhead powerlines by presenting theory, algorithms, and a methodology for implementation in a computer program. The development of a computer program by object-oriented modeling and expert system rules is also described in detail. The end product is easy to use and suitable for application in all geographic regions. The different methods of increasing line ampacity by FACTS are described, and the impact of higher transmission line ampacity on electric and magnetic fields is analyzed with numerical examples. Application of the powerline ampacity system in the economic operation of a power system is presented, and considerable cost savings are shown by the deferment of capital investment required for the construction of new lines, and by enabling greater utilization of low-cost energy sources.

1.3 ELECTRIC POWER SYSTEM OVERVIEW

1.3.1 Transmission Grid

The electric power system is comprised of an interconnected transmission grid that is used to connect diverse generation sources for the distribution of electricity to load centers in the most economical manner. Figure 1.1 shows the transmission grid

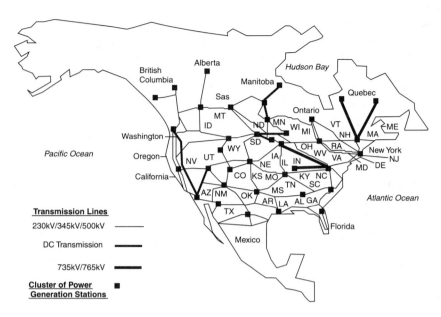

FIGURE 1.1 North American Transmission grid.

of North America with transmission lines having voltages 230 kV and above. The total length of transmission lines at each voltage category is shown in the Figure 1.2. The installed generation capacity is approximately 750 GW, which is expected to grow at the rate of 2% per year. Assuming an annual load factor of 0.5, approximately 3200 billion kilowatt-hours will be distributed through the transmission network in the year 2000. Figures 1.3–1.5 show electricity production in the U.S. and the world.

According to the U.S. Department of Energy,* about 10,000 circuit km of transmission lines are planned to be added by the year 2004. The total cost of adding new transmission lines is approximately three billion dollars. In addition to the high cost of adding new transmission lines, environmental factors related to land use and EMF are also required to be considered before the construction of new lines. Dynamic rating of transmission lines offers substantial cost savings by increasing the capacity of existing lines such that the construction of new lines may be postponed in many cases.

1.3.2 OVERHEAD TRANSMISSION LINE

The overhead transmission line consists of towers, conductors, insulators, and line hardware for the jointing of conductors and for properly supporting the high-voltage line to the transmission line tower. The most common type of transmission line

---

* Arthur H. Fuldner, *Upgrading Transmission Capacity for Wholesale Electric Power Trade*, U.S. Department of Energy publication on the World Wide Web, December 30, 1998.

Introduction

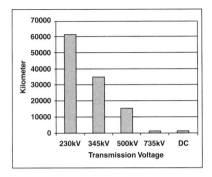

FIGURE 1.2 Total transmission line circuit km in North America according to transmission voltage category.

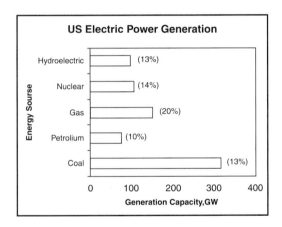

FIGURE 1.3 U.S. electric utility generation capacity.

towers are self-supporting towers, and guyed and pole towers. Some typical examples of towers are shown in Figures 1.6–1.9.

Aluminum Conductor Steel Reinforced (ACSR) is the most widely used type of current-carrying conductor. All Aluminum Conductors (AAC) are used in coastal regions for high corrosion resistance and also for applications requiring lower resistance, where the high strength of a steel core is not required. More recently, All Aluminum Alloy conductors have been used for their light weight and high strength-to-weight ratio, which enables longer spans with less sag. Other hybrid conductors having various proportions of aluminum, aluminum alloy, and steel wires are also used for special applications. The popular type of hybrid conductors are Aluminum Conductor Alloy Reinforced (ACAR) and Aluminum Alloy Conductor Steel Reinforced (AACSR). Some examples of commonly used powerline conductors according to various standards are given in Appendix A (Thrash, 1999; Koch, 1999; Hitachi, 1999).

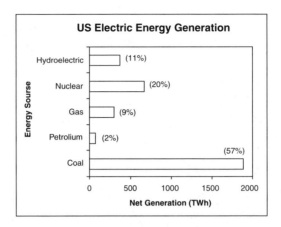

FIGURE 1.4 U.S. utility electric energy production.

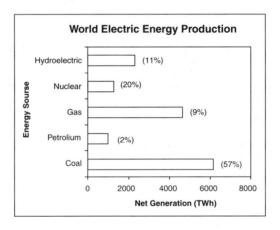

FIGURE 1.5 World electric energy production.

FIGURE 1.6 Self supporting tower.

Introduction

FIGURE 1.7 Guyed tower.

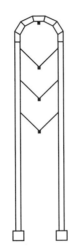

FIGURE 1.8 Ornamental tower.

FIGURE 1.9 Tubular tower.

High-temperature conductors are used for bulk power transmission in heavily loaded circuits where a high degree of reliability is required. Steel Supported Aluminum Conductor (SSAC) allows high-temperature operation with minimum sag. In the SSAC conductor, the current-carrying aluminum wires are in the annealed state and do not bear any tension. The tension is borne entirely by the high-strength steel wires. In the newer high-temperature, high-ampacity conductors, aluminum zirconium alloy wires are used to carry high current, and Invar alloy reinforced steel wires are used for the core.

Recently, compact conductor designs have been available that offer lower losses for the same cross-sectional area of the conductor. Compact design is made possible by the trapezoidal shaping of wires instead of wires having the circular cross-sections used in conventional ACSR conductors. For better aerodynamic performance, conductors are also available with concentric gaps inside the conductor which offer better damping of wind-induced vibrations.

Another recent development in transmission line conductor technology is the integration of optical fiber communication technology in the manufacture of powerline conductors. In an Optical Ground Wire (OPGW) system, a fiberoptic cable is placed inside the core of the overhead ground wire. In certain transmission line applications, the fiberoptic cable is placed inside the core of the power conductor. Communication by fiber optics offers a noise-free system of data communication in the electric utility environment since communication by optical fiber is unaffected by electromagnetic disturbances. The different types of conductors are shown in Figure 1.10. Important physical properties of the different types of wires used in the manufacture of powerline conductors are given in Appendix B.

1.3.3 HIGH-VOLTAGE SUBSTATION

The electric substation is an important component of the electric power system. The substation is a hub for receiving electricity from where electricity is distributed to load centers, as well as to other substations. Voltage transformation is carried out in the substation by transformers. A transmission substation is generally used for interconnection with other substations where power can be rerouted by switching action. In a distribution substation, electricity is received by high-voltage transmission lines and transformed for distribution at lower voltages. Besides transformers, there are other important devices in a substation, including bus bars, circuit breakers, interrupters, isolators, wave traps, instrument transformers for the measurement of high voltage and current, inductive and capacitive reactors for the control of reactive power flow, protective relays, metering, control and communication equipment, and other low-voltage equipment for station auxiliary power supply.

A typical layout of a high-voltage substation is shown in Figure 1.11. When a dynamic line rating system is implemented in an electric utility system, it is also important to have knowledge of the current rating of all substation equipment in addition to powerline conductor ratings. Substation switching devices are generally designed to withstand short circuit currents, and have sufficient continuous overload current capability. Transformer ratings, on the other hand, need to be examined more closely. A system of dynamic rating of substation equipment may be implemented

Introduction

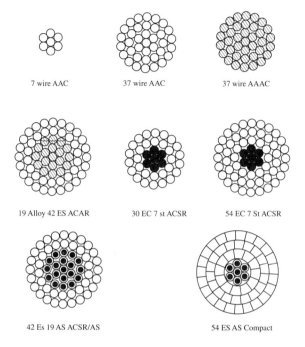

FIGURE 1.10 Transmission line conductors.

by real-time monitoring of equipment temperature by installing sensors, or by inferring equipment temperature by the measurement of current flowing through the device and monitoring weather conditions at the location of the substation.

1.3.4 ENERGY CONTROL CENTER

The electric power system comprising generation stations, transmission and distribution lines, and substations is controlled by a system of energy control centers. Each electric power company operates its electricity supply system in a given geographic region through one or more energy control centers, as shown in Figure 1.13. For example, Electricité de France (EDF), the national electric power supply company of France, operates its electric power supply system through one central control center and seven regional control centers. Control centers are responsible for the control of power generation, load forecasting, performing load flow, dynamic and transient stability analysis, contingency analysis, and switching operations in the substations. Control centers constantly monitor the condition of all transmission lines and substations in their respective regions, and, in the event of a failure of a component in the network, control actions are taken to remedy the problem.

Modern control centers are operated through a network of computers having intelligent programs called "expert systems." These expert systems perform a variety of tasks from energy management to alarm processing and fault diagnosis, providing assistance to control system operators for better decision making, which is especially

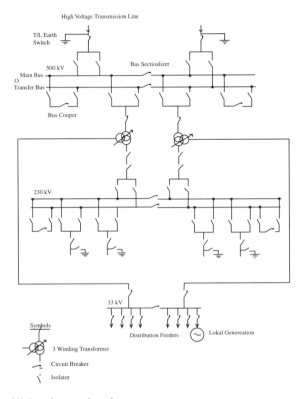

FIGURE 1.11 High voltage substation.

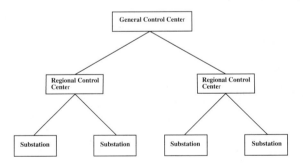

FIGURE 1.12 Typical power system hierarchy.

useful during an emergency. The powerline ampacity system described in this book is an expert system for the evaluation of transmission line ampacity, which is expected to be an integral part of a modern energy control system.

Courtesy National Grid Company

FIGURE 1.13 Energy Control Center.

1.4 FACTORS AFFECTING TRANSMISSION CAPACITY AND REMEDIAL MEASURES

The effects of elevated temperature operation are loss of tensile of conductors and permanent elongation of conductors. The loss of strength model is given by Harvey (Harvey, 1972) and (Morgan, 1978). The models for permanent elongation of conductor is given in a Cigré 1978 report. A recursive estimation algorithm for calculating the loss of tensile strength and permanent elongation due to heating in service from the probability distribution of conductor temperature is described by the author (Deb et al., 1985). A study for the assessment of thermal deterioration of transmission line conductor from conductor temperature distribution was presented recently by Mizuno et al. (1998). The results were presented by the author (Deb, 1993) for practical line operating conditions.

The remedial measures that are proposed to reduce the possibility of transmission line conductor overheating comprise the use of line ampacity programs and the monitoring of transmission line current and/or temperature.

Special conductors may be used to transfer higher currents in highly congested transmission circuits. A recent study conducted by the author and KEPCO* (Wook, Choi, and Deb, 1997) shows that transmission capacity may be doubled by the application of new types of powerline conductors. The new types of conductors are capable of operating at significantly higher temperatures with less sag and without any thermal deterioration. There is general agreement that transmission line magnetic fields** have minimum impact on the environment, and there are no harmful effects of magnetic fields on human beings. A recent research study conducted by EPRI (Rashkes and Lordan, 1998) presents new transmission line design considerations to lower magnetic fields. This study is important from the

---

* Author worked as a consultant for Korea Electric Power Company (KEPCO), South Korea.
** EMF Conference. National Academy of Science, U.S.A. 1994, concluded that there are no harmful effects due to powerline electric and magnetic fields.

point of view of transmission line ampacity so that future transmission lines can be constructed with higher power transfer capability and minimum magnetic field.

1.5 NEW DEVELOPMENTS FOR TRANSMISSION CAPACITY ENHANCEMENT

There are other electrical network constraints that must be satisfied before transmission lines can be operated at their maximum thermal capacities. The most important constraints are voltage levels and generator stability limits. New methods and devices to improve transmission system voltage levels and generation stability limits include FACTS (Flexible AC Transmission System) (Hingorani, 1995).

FACTS technology makes use of recent developments in modern power electronics and superconductivity (Feak, 1997) to enhance transmission capacity. A recent FACTS development is the Unified Power Flow Controller (UPFC) (Norozian et al., 1997). Another important development is the invention of a new type of generator called the "Powerformer" (*MPS Review*, 1998b) that eliminates the need for a transformer by generating electricity at high voltage at the level of transmission system voltage. The new type of generator produces greater reactive power to elevate grid voltage levels, and also enhances generation stability when required. Therefore, by connecting Powerformer directly to the transmission grid, yet higher levels of transmission capacity may be achieved. These studies show that there is considerable interest in maximizing the capacity of existing assets. By the introduction of these new technologies in the electrical power system, it is now becoming possible to operate transmission lines close to thermal ratings, when required.

1.6 DYNAMIC LINE RATING COST-BENEFIT ANALYSIS

A cost benefit analysis was carried out by the cost capitalization method and the results are presented in Table 1.1. It is assumed that line current will increase at the rate of 2.5% per year. The results show that the capitalized cost of higher losses due to the increase in line current by deferment of new line construction for a period of 10 years is significantly lower than the cost of constructing a new line in the San Francisco Bay area.

In addition to cost savings achieved by postponing the construction of new lines, dynamic line rating systems also offer substantial operational cost savings. In Chapter 10, a study is presented which show 16% economy achieved by dynamic line rating by facilitating the transfer of low-cost surplus hydroelectric energy through overhead lines. To undertake this study, an economic load flow program was developed to simulate an interconnected transmission network with diverse generation sources (Hall and Deb, 1988a; Deb, 1994; Yalcinov and Short, 1998).

1.7 CHAPTER SUMMARY

An introduction to the subject of transmission line ampacity is presented in this chapter by giving an overview of the electric power system. The significance of the study and the main contributions in each chapter are summarized.

TABLE 1.1
Cost–Benefit Analysis

Year	Line Current, Increase @ 2.5%/yr, A	Annual Energy Loss Increase, @ 10c/kWh, $	Present Value of Annual Loss, @ 10%/yr Interest, $
1995	800	0	0
1996	820	2,759	2,508
1997	841	5,657	4,675
1998	862	8,702	6,538
1999	883	11,902	8,129
2000	905	15,263	9,477
2001	928	18,794	10,609
2002	951	22,505	11,548
2003	975	26,403	12,317
2004	1000	30,498	12,934
		Total present value of loss, $/mile	78,736
		Cost of new line, $/circuit/mile	200,000
		Saving by LINEAMPS, $/circuit/mile	121,264

Note: The following assumptions are made in the above calculations:
Line load loss factor = 0.3, based on system load factor = 0.5.
Conductor is ACSR Cardinal, 1 conductor/phase, single circuit line.
Rate of interest is 10% /year.
Line load increase by 2.5% /year.
Static normal ampacity of the line is 800 A.

As the demand for electricity grows, new methods and systems are required to maximize the utilization of existing power system assets. High-voltage transmission lines are critical components of the electric power system. Due to environmental, regulatory, and economical reasons it is not always possible to construct new lines, and new methods are required to maximize their utilization. The object of this book is to present a study of transmission line conductor thermal modeling, to develop a methodology for the rating of transmission lines for implementation in a computer program that is suitable for all geographic regions, and to present the applications of line ampacity in the operation of electric power systems.

The development of a complete line ampacity system having transmission line, weather, and conductor models that can be easily implemented in all geographic regions was a major challenge. For this reason it was necessary to develop a computer program that will adapt to different line operating standards followed by power companies in the different regions of the world. This was accomplished by developing an expert system and object-oriented modeling of the line ampacity system.

The economic incentives for implementing a dynamic line rating system are clearly established by showing the approximately 60% cost saving by the deferment of new line construction. The factors limiting line capacity are clearly brought out, and the means to overcome these are explained.

2 Line Rating Methods

2.1 HISTORICAL BACKROUND

2.1.1 Early Works on Conductor Thermal Rating

Faraday was one of the early researchers who conducted theoretical and experimental research to study the heating of wires by electric current (Faraday, 1834).* Some early works on transmission line conductor thermal rating were conducted in France (Legrand, 1945) that realized the importance of transmission line conductor thermal ratings. A transmission line rating system using temperature monitoring by a thermal image of conductors was developed in Belgium (Renchon, 1956).

A steady-state ampacity model based on the conductor heat balance equation was presented in 1956 (House and Tuttle, 1956). For the short-term rating of transmission line conductors, Davidson (1969) presented a solution to the differential equation of conductor temperature by using the Eulers method. All of the above research shows that there has long been considerable interest in maximizing the transmission capacity of overhead lines. Weather modeling for transmission line ampacity was first presented by the author at the Cigré Symposium** on High Currents (Deb et al., 1985).

2.1.2 IEEE and Cigré Standards

IEEE (IEEE Standard 738, 1993) and Cigré (Cigré, 1992, 1997, 1999) offer standard methods for the calculation of transmission line ampacity in the steady, dynamic, and transient states. The Cigré report presents a three-dimensional thermal model of conductors for unsteady-state calculation. A similar model was presented at the IEEE (Hall, Savoullis, and Deb, 1988) for the calculation of thermal gradient of conductor from surface to core.

2.1.3 Utility Practice

Electric power companies*** generally assume that the ampacity of transmission line conductors is constant. Ampacity calculations are commonly based upon the following conservative assumptions of ambient temperature, wind speed, solar radiation, and maximum conductor temperature:

---

* Michael Faraday, Electricity, Encyclopaedia Britannica, Great Books # 42, page 686.
** Cigré Symposium: High Currents in Power Systems under Normal, Emergency and Fault Conditions, Brussels, Belgium, 3–5 June, 1985, devoted to the subject of transmission line ampacity.
*** Electricité de France, Paris, is the national electric power company of France (Urbain, 1998); Pacific Gas & Electric Co. San Francisco, CA, (PG&E Standard 1978); Central Board of Irrigation and Power, India (Deb et al., 1985).

- Ambient temperature = 40°C
- Wind speed = 0.61 m/s (2 ft/s)
- Solar radiation =1000 W/m2
- Maximum conductor temperature = 80°C

It is well known that weather conditions are never constant. Therefore, during favorable weather conditions when ambient temperature is lower than the assumed maximum or wind speed is higher than the assumed minimum, or during cloudy sky conditions, higher ampacity is possible without exceeding the allowable maximum temperature of the powerline conductor. For the above reasons, many utilities have started adapting line ratings to actual weather conditions to increase line capacity. A dynamic line rating system can provide further increase in line ampacity for short durations by taking into consideration the heat-storage capacity of conductors.

2.2 LINE RATING METHODS

2.2.1 Defining the Line Ampacity Problem

The problem of determining the thermal rating of an overhead powerline can be stated as follows: based on existing and forecast weather conditions at several locations along the transmission line route, determine the maximum current that can be passed through the line at a given time (t) such that the conductor temperature (Tc) at any section of the line does not exceed the design maximum temperature (Tmax) of the line.

Stated formally,

$$\forall_l I_{l,t} = \min(I_{l,j,t}) \tag{2.1}$$

$$I_{l,j,t} = f(Ws_{k,j,t}, Wd_{k,j,t}, Ta_{k,j,t}, Sr_{k,j,t}, Tc_{l,j}, C_{l,j}, D_{l,j}) \tag{2.2}$$

$$Tc_{l,j} \leq Tmax_{l,j} \tag{2.3}$$

Where,

I = Ampacity (Ampere)
Ws = Wind speed
Wd = Wind direction
Ta = Ambient temperature
Sr = Solar radiation
Tc = Conductor temperature
C = Conductor
D = Direction of line
l = 1,2,3… L transmission lines
j = 1,2,3… J line sections
t = 1,2,3… T time (T = 168 h in LINEAMPS)
k = 1,2,3… K weather stations

Line Rating Methods 17

The LINEAMPS computer program described in this book finds a solution to the above line ampacity problem.

2.2.2 STATIC AND DYNAMIC LINE RATINGS

Transmission line rating methods are broadly classified into two categories: static and dynamic line rating. The static line rating system is widely used because of its simplicity, as it does not require monitoring weather conditions or installation of sensors on the transmission line conductor. The static rating of transmission lines in a region is generally determined by analysis of historical weather data of that region for the different types of conductors used in the transmission lines. Generally, static line ratings are fixed for a particular season of the year, and many electric power utilities have different line ratings for summer and winter. For example, the static line rating of some typical conductor sizes used by PG&E in the region of the San Francisco Bay area is given in Table 2.1.

TABLE 2.1
Ampacity of ACSR Conductors

Conductor Size, mm^2	Summer Coastal		Winter Coastal	
	Normal	Emergency	Normal	Emergency
210	382	482	550	616
264	442	558	640	716
375	550	697	801	898
624	752	959	1108	1243
749	919	1133	1218	1393
874	1060	1312	1448	1625
1454	1319	1642	1814	2040

Basis for Table 2.1
Summer ambient temperature = 37°C with sun
Winter ambient temperature = 16°C without sun
Wind velocity = 0.6 m/s perpendicular to conductor axis
Conductor temperature, normal condition = 80°C
Conductor temperature, emergency condition A = 90°C (100 hr total)
Conductor temperature, emergency condition B = 100°C (100 hr total)
(Emergency B ratings are shown in the table)
Emissivity = 0.5
Conductivity of aluminum = 61% IACS

Dynamic line ratings are obtained by online or offline methods. Online line rating methods include monitoring conductor temperature or tension, and weather conditions all along the transmission line route. Conductor temperature is monitored by installing conductor temperature sensors at certain sections of the transmission line. Conductor tension is overseen by tension monitors that are attached to insulators on tension towers. Unlike temperature monitoring systems, tension monitors are required to be located only at anchor towers. In both monitoring systems, sensor data is communicated to a base station computer by a radio communication device installed on the sensor, and the ampacity of the line is calculated at the base station computer from this data.

In the offline system, line ratings are obtained uniquely by monitoring weather conditions along the transmission line route. An offline system may also include monitoring conductor sag by pointing a laser beam at the lowest point of the conductor in a span. The ampacity of a line is calculated from conductor sag and weather data by taking a series of measurements of conductor sag at different transmission line spans along the length of the line.

2.2.3 WEATHER-DEPENDENT SYSTEMS

Weather-dependent line rating systems were proposed by several researchers (Cibulka et al., 1992; Douglass, 1986; Hall and Deb, 1988b; Mauldin et al., 1988; Steeley et al., 1991). These methods require weather data on a continuous basis. Diurnal weather patterns of the region are considered for the prediction of line ampacity several hours in advance. The existence of daily and seasonal cyclical weather patterns are well known, and their usefulness to forecast powerline ampacity was recognized by many researchers (Foss and Maraio, 1989), (Hall and Deb, 1988b).

A weather-dependent line rating system developed in the UK is described in a Cigré article (Jackson and Price, 1985). Similarly, a weather-dependent real-time line rating system has been developed for the Spanish 400 kV transmission network (Soto, et al., 1998). In the Spanish system, real-time measurements of wind speed, wind direction, ambient temperature, and solar radiation from several weather stations are entered into a computer where a line ampacity program calculates steady-state and dynamic ampacity. Foss and Maraio (1989) described a line ampacity system for the power system operating environment. They were also interested in forecasting transmission line ampacity. In their method, line ampacity is adjusted based on previous 24-hour weather data. Because of these assumptions, the accuracy of the system in forecasting transmission line ampacity several hours ahead is somewhat limited.

In the LINEAMPS program (Deb, 1995a, 1995b), the periodic cyclical pattern of wind speed and ambient temperature are considered in a unique manner to forecast powerline ampacity. Weather patterns of a region are stored in Fourier series in each weather station object. A method in each weather station object generates hourly values of meteorological data from this series. The powerline objects have a plurality of virtual weather sites that receive their data from a plurality of weather station objects, and a method in each powerline object determines the minimum hourly values of line ampacity up to seven days in advance. The number of virtual weather stations that can be accommodated in a powerline is limited only by the computer processing speed and memory, whereas installing an unlimited number of temperature sensors on a transmission line is not economical. Due to these reasons, a

Line Rating Methods

weather-dependent line rating system is expected to be more reliable and more accurate than systems utilizing real-time measurements from a limited number of locations.

2.2.4 ONLINE TEMPERATURE MONITORING SYSTEM

U.S. Patent 5140257 (system for rating electric power transmission lines and equipment, 1992) was given for a transmission line rating system that calculates the current carrying capacity of one or more powerlines by the measurement of conductor temperature and meteorological conditions on the line. In this method, line ampacity is calculated by the measurement of conductor temperature and by the solution of the conductor heat balance equation as follows:

$$I = \sqrt{\frac{P_r + P_c - P_s}{R_{ac}}} \qquad (2.4)$$

where,

P_r = Heat lost by radiation, W/m
P_c = Heat lost by convection due to the cooling effect of wind, W/m
P_s = Heat gained by solar radiation, W/m
R_{ac} = AC resistance of conductor, ohm/m

In the above equation, P_r, P_c, P_s, and R_{ac} are functions of conductor temperature. An on-line temperature monitoring system using Power Donut ™ temperature sensors is shown in Figure 2.1. The Power Donut ™ temperature sensor is shown in Figure 2.2.

Davis's system (Davis, 1977) required the installation of conductor temperature sensors as well as meteorological sensors at several locations along powerlines. Real-time conductor temperature, meteorological data, and line current are continuous input to a computer system where line ampacity is calculated. The computer system requires special hardware and software for data acquisition from remote sensor locations via special telecommunication networks. The online monitoring systems described in the IEEE and Cigré papers (Davis, 1977; Howington and Ramon, 1984; Renchon and Daumerie, 1956) are not widely used because of transmission distance, communication requirements, and maintenance problems. The new line ampacity system does not require real-time continuous input of meteorological data, line current, or conductor temperature measurements from the powerline. Powerline and conductor ampacity is estimated by the program from user input and by synthetic generation of weather data from self-generating weather station objects. General purpose weather forecast data available from the internet are used in the LINEAMPS program.

A real-time dynamic line-rating model was proposed (Black and Byrd, 1983). In this method, line ampacity is predicted accurately by real-time numerical solution of the following conductor temperature differential equation at a location:

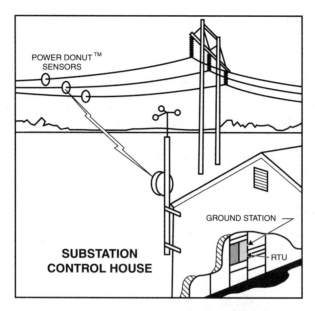

FIGURE 2.1 On-line temperature monitoring system is comprised of Power Donut™ temperature sensors, and weather station and ground station RTU. (© Courtesy Nitech, Inc.)

$$M \cdot c_p \frac{dT_{av}}{dt} = P_j + P_s + P_m - P_r - P_c \quad (2.5)$$

Where,

$M = \gamma \cdot A$, conductor mass, kg/m
A = conductor area, m²
T_{av} = average conductor temperature, °C
γ = conduct or density, kg/m³

$$T_{av} = \frac{T_c + T_s}{2} \quad (2.6)$$

T_c = Conductor core temperature, °C
T_s = Conductor surface temperature, °C measured by line temperature sensor
P_r = Heat lost by radiation, W/m
P_c = Heat lost by convection due to the cooling effect of wind, W/m
P_s = Heat gained by solar radiation, W/m
P_m = Magnetic heating, W/m
P_j = Joule heating, W/m

The following expression for the calculation of real-time dynamic ampacity was obtained by the author:

Line Rating Methods

FIGURE 2.2 Power Donut™ temperature sensor. (© Courtesy Nitech, Inc.)

$$I = \sqrt{\frac{\{T_{max} - T_{initial} \exp(-\Delta t/\tau)\}}{C_1\{1 - \exp(-\Delta t/\tau)\}} - C_2} \quad (2.7)$$

T_{max} = Max conductor temperature
$T_{initial}$ = Initial temperature $T_{initial}$, and time Δt
C_1, C_2 = Constants

The different terms in the above equation are described in Chapter 3.

The calculation of dynamic ampacity by the above equation requires real-time conductor temperature and meteorological data on a continuous basis.

2.2.5 Online Tension Monitoring System

The online tension monitoring system is used to predict transmission line ampacity by measurement of conductor tension at tension towers along the transmission line (Seppa, et al., 1998). Since conductor tension is a function of conductor temperature, the ampacity of the transmission can be obtained by real-time monitoring of conductor tension as follows.

$$\frac{\sigma_2}{E} - \frac{(\varpi \cdot L)^2}{24\sigma_2^2} + \alpha(Tc_2 - Tc_1) + \Delta Ec = \frac{\sigma_1}{E} - \frac{(\varpi \cdot L)^2}{24\sigma_1^2} \quad (2.8)$$

σ_1, σ_2 = stress at state1 and state2, respectively, kg/mm²
Tc_1, Tc_2 = conductor temperature at state1 and state2, °C
E = Young's modulus of elasticity, kg/mm²
ϖ = specific weight of conductor, kg/m/mm²
L = span length, m
ΔEc = inelastic elongation (creep) mm/mm
α = coefficient of linear expansion of conductor, °C^{-1}

Therefore, by measurement of conductor tension and by knowledge of initial conditions, the temperature of the conductor is obtained by the solution of the above equation. Transmission line ampacity is then calculated by the solution of conductor heat balance, Equation 2.5. This method of monitoring a transmission line has the added advantage of monitoring ice-loads as well. The major disadvantages of this method are that it requires taking the transmission line out of service for installation and maintenance. It may be feasible to install such devices on certain heavily loaded lines, but is impractical and expensive to install tension monitors on all transmission and distribution lines for line ampacity predictions of all overhead lines in a system.

2.2.6 Sag-Monitoring System

This is an offline method of real-time line rating by monitoring conductor sag. It is an offline method because it does not require the installation of any device on the transmission line conductor. Therefore, this system does not require taking the line out of service during installation or maintenance of the sag-monitoring device. In this method, conductor sag is measured by pointing a laser beam at the lowest point of the conductor in a span. Ampacity is calculated from conductor sag and by measurement of weather conditions. The ampacity of the transmission line is then obtained by taking a series of measurements at different transmission line spans along the length of the line.

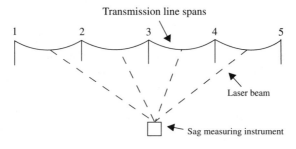

FIGURE 2.3 Sag-monitoring line ampacity system.

Conductor sag is calculated approximately by the well-known parabola equation:

$$\text{Sag} = \frac{WL^2}{8T} \qquad (2.9)$$

W = conductor weight, kg/m
T = conductor tension, kg

Having calculated conductor tension T from (2.9), the temperature of the conductor is then obtained from (2.8), and then ampacity is calculated from (2.5).

2.2.7 Distributed Temperature Sensor System

With the development of the powerline communication system by a fiberoptic cable integrated with a powerline conductor, it is now possible to have a distributed system of fiberoptic conductor temperature sensors that will span the entire length of the transmission line. A distributed temperature sensor system will result in a more accurate real-time line rating system, since a fiberoptic cable will be used for data transmission and will eliminate the need for a separate communication system for transmission of conductor temperature data from a transmission line to utility power control center. At the present time, the fiberoptic cable is embedded inside the core of powerline ground wire. This kind of conductor, with a fiberoptic cable in the core, is called an OPGW conductor. A fiberoptic cable also may be placed within the core of a phase conductor in high-voltage lines. In low-voltage distribution lines, the fiberoptic cable may be wrapped over the conductor. An example of an OPGW conductor on overhead line is shown in Figure 2.4.

Probabilistic ratings are used by several power companies (Deb et al., 1985; Giacomo, Nicolini, and Paoli, 1979; Koval and Billinton, 1970; Urbain, 1998). Probabilistic ratings are determined by Monte Carlo simulation of meteorological variables, and by the solution of the conductor heat balance equation. From the resultant probability distribution of conductor ampacity, line ratings are determined. An attractive feature of this technique is that continuous input of real-time weather data is not required. A limitation is that line ampacity is not adaptive to real weather conditions. However, it must be mentioned that probability modeling of conductor temperature is useful for the prediction of conductor performance in service.

Probability modeling of conductor temperature is used to predict the loss of tensile strength and permanent elongation of conductor during the lifetime of the transmission line conductor (Deb, 1985, 1993; Hall and Deb, 1988b; Mizuno et al., 1998). The author has obtained the probability distribution of conductor temperature by Monte Carlo simulation of time series stochastic models of the meteorological variables and transmission line current. By using time series stochastic models, it is possible to consider the correlation between the different variables (Douglass, 1986). For example, it is well known that electricity demand depends upon weather conditions.

FIGURE 2.4 Power line conductors with fiberoptic cable.

A time series stochastic and deterministic model was used to predict real-time probabilistic ratings of transmission line ampacity up to 24-hours in advance based on ambient temperature measurements only, and by assuming constant wind speed (Steeley et al., 1991). A stochastic model was also used to forecast solar radiation (Mauldin et al., 1991) and wind speed (Hall and Deb, 1988b). The general form of the stochastic model is given below:

$$Ta(t) = A + A_2 \cdot Sin(wt) + A_3 \cdot Sin(2wt) \\ + A_4 \cdot Cos(wt) + A_5 \cdot Cos(2wt) \\ + A_6 \cdot Z(t-1) + A_7 \cdot Z(t-2) \qquad (2.10)$$

$Z(t-1)$, $Z(t-2)$ = difference of measured and predicted temperature at time $(t-1)$ and $(t-2)$ respectively
$A_1, A_2, A_3, A_4, A_5, A_6, A_7$ are the coefficients of the model
$\omega = 2\pi/T$ = fundamental frequency
$T = 24$ hour = period

In the LINEAMPS program, Fourier series models of ambient temperature and wind speed are used to generate weather data. Because weather patterns are stored in weather objects, it eliminates the need for real-time measurements on a continuous basis. Weather data is required only when weather conditions change. One of the limitations of the stochastic model is that it is unsuitable for the predictions of hourly values of ambient temperature for more than 24 hours in advance. This is to be expected, because time-series models are statistical models that do not consider a physical model of the atmosphere. In this regard, National Weather Service forecasts are generally more accurate for long-term weather predictions because they are derived from atmospheric models. Therefore, the parameters of the weather models in the LINEAMPS program are adjusted to National Weather Service forecasts.

2.2.8 OBJECT-ORIENTED MODELING AND EXPERT LINE RATING SYSTEM

The estimation of powerline ampacity by the application of object-oriented modeling and expert system rules was first presented by the author (Deb, 1995). It was shown for the first time how object-oriented modeling of transmission line ampacity enabled program users to easily create new lines and new conductors, and for weather stations to estimate line ampacity up to seven days in advance. It was shown that powerline objects not only have methods to predict line ampacity, but are also convenient repositories for the storage of line data and ampacity that are easily retrieved and displayed on a computer screen.

Important electric power companies are embracing the object-model approach to meet their information requirements for the year 2000 and beyond (*MPS Review* article, 1998). For example, EDF has selected object-model technology for the management and operation of the transmission grid in France. Expert systems (Kennedy, 2000) are developed in the electric power system for power quality machine diagnosis, alarm-processing (Taylor et al., 1998) and for power system fault analysis (Negnevitsky, M., 1998). LINEAMPS is the first expert system for the estimation of transmission line ampacity (Deb, Anjan K., 1995).

Integrated Line Ampacity System

LINEAMPS (Deb, 1995a, 1995b), (Wook et al., 1997) is an integrated line ampacity system having a transmission line model, conductor model, and weather model to forecast line ampacity up to seven days in advance. It has provision for steady-state rating, dynamic line rating, and transient rating. The concept of steady, dynamic, and transient line ratings are used in this program for the first time. A direct solution of the conductor temperature differential equation is used, and an analytical expression for the direct solution of dynamic line ampacity is presented for the first time (Deb,1998a). A three-dimensional conductor thermal model is used to calculate conductor thermal gradient (Deb, 1998a, 1998b).

An important contribution of the new line ampacity system is the ability to self-generate hourly values of weather data from statistical and analytical models, eliminating the need for real-time weather data on a continuous basis. The author previously developed an algorithm (Deb, 1993) for synthetic generation of weather data by time-series analysis and recursive estimation. The idea of self-generation, synthetic generation, or artificial generation of meteorological data by a Fourier series weather model of ambient temperature and wind speed of a region evolved from these developments. Synthetic generation of weather data from a model is also useful to evaluate the probability distribution of transmission line conductor temperature in service (Hall, Deb, 1988b). The probability distribution of transmission line conductor temperature is required to calculate the thermal deterioration of transmission line conductor (Mizuno et al., 1998, 2000).

2.3 CHAPTER SUMMARY

Transmission line rating methods are introduced by presenting a critical review of literature, from the early works on conductor thermal rating to modern applications of object-oriented modeling, expert systems, recursive estimation, and real-time transmission line ratings. The line ampacity problem is clearly defined, and the various methods of rating power transmission lines are critically examined to show the advantages and deficiencies of each method. The various methods of rating transmission lines includes static and dynamic thermal ratings, weather-dependent system, temperature monitoring system, tension monitoring system, sag monitoring, distributed fiberoptic sensors, and probabilistic rating methods.

The development of an integrated powerline ampacity system having transmission line, weather, and conductor thermal models that can be easily implemented in all geographic regions was a major challenge. For this reason, it was necessary to develop a computer program that will adapt to the different line operating standards followed by power companies in the different regions of the world. This was accomplished by object-oriented modeling and by developing an expert system computer program called LINEAMPS

3 Theory of Transmission Line Ampacity

3.1 INTRODUCTION

As mentioned in the previous chapters, the current-carrying capacity of a transmission line conductor is not constant but varies with weather conditions, conductor temperature, and operating conditions. Line ampacity is generally based on a maximum value of conductor temperature determined by the type of conductor, and depends upon the following operating conditions:

- Steady-state
- Dynamic state
- Transient state

The conductor is assumed to be in the steady-state during normal operating conditions when the heat gained due to line current and solar radiation equals the heat lost by cooling due to wind and radiation. During steady-state conditions, the transmission line current is considered constant, weather conditions are assumed stable, and the temperature of the conductor is fairly uniform.

Dynamic conditions arise when there is a step change in line current. Line energization or sudden changes in line current due to a failure on one circuit are examples of dynamic operating conditions. A typical example of dynamic loading is when the load from the faulted circuit in a double circuit line is transferred to the healthy circuit. Due to the thermal inertia of the conductor, short-term overloads may be supplied through the line without overheating the conductor before steady-state conditions are reached.

Transient conditions arise due to short-circuit or lightning current. During transient conditions there is no heat exchange with the exterior, and adiabatic conditions are assumed.

In the following section, the equations for the calculation of conductor temperature and ampacity are derived from the general heat equation for steady-state, dynamic state, and transient conditions. Then, equations for the radial conductor temperature differential from surface to core of a conductor are developed from the same equations.

This chapter prepares the framework for computer modeling of the line ampacity system described in Chapter 8.

3.2 CONDUCTOR THERMAL MODELING

3.2.1 GENERAL HEAT EQUATION

Starting with the general heat equation of the transmission line conductor, the solution of the transmission line ampacity is found for each of the above operating conditions as follows:

The general heat equation* for a transmission line conductor is given by,

$$\nabla^2 T + \frac{q}{k} = \frac{1}{\alpha} \frac{\partial T}{\partial t} \tag{3.1}$$

Where,

$$\nabla^2 = \text{Laplacian operator,}$$

$$\nabla^2 = \frac{\partial^2}{\partial x^2} + \frac{\partial^2}{\partial y^2} + \frac{\partial^2}{\partial z^2}$$

In cylindrical coordinates,

$$\nabla^2 T = \frac{\partial^2 T}{\partial r^2} + \frac{1}{r} \frac{\partial T}{\partial r} + \frac{1}{r^2} \frac{\partial^2 T}{\partial \varphi^2} + \frac{\partial^2 T}{\partial z^2} \tag{3.2}$$

T = conductor temperature
r = radial length
φ = azimuth angle
z = axial length
q = power per unit volume
k = thermal conductivity of conductor
α = thermal diffusivity given by,

$$\alpha = \frac{k}{\gamma c_p}$$

c_p = specific heat capacity
γ = mass density

From (3.1) and (3.2) we obtain,

$$\frac{\partial T}{\partial t} = \frac{\lambda}{\gamma \cdot c_p} \left(\frac{\partial^2 T}{\partial r^2} + \frac{1}{r} \frac{\partial T}{\partial r} \right) + q \tag{3.3}$$

λ = Thermal conductivity

---

* J.F. Hall, A.K. Deb, J. Savoullis, Wind Tunnel Studies of Transmission Line Conductor, IEEE, *Transactions on Power Delivery*, Volume 3, Number 2, April 1988.

Theory of Transmission Line Ampacity

Equation 3.3 may be solved numerically with appropriate initial and boundary condition,[1,2,3] or solved analytically by making certain simplifying assumptions that are presented in the following section.

3.2.2 Differential Equation of Conductor Temperature

For practical consideration of transmission line conductor heating, it is possible to make the following assumption with sufficient accuracy:*

$$T_{av} = \frac{T_c + T_s}{2} \tag{3.4}$$

Where,

T_{av} = average conductor temperature
T_c = conductor core temperature
T_s = conductor surface temperature

With the above assumption, we can calculate the average conductor temperature by the solution of the following differential equation obtained from (3.3):

$$M \cdot c_p \frac{dT_{av}}{dt} = P_j + P_s + P_m - P_r - P_c \tag{3.5}$$

Where,

$M = \gamma \cdot A$, conductor mass, kg/m
A = conductor area, m^2
P_j = joule heating, W/m
P_s = solar heating, W/m
P_m = magnetic heating, W/m
P_r = radiation heat loss, W/m
P_c = convection heat loss, W/m

3.2.3 Steady-State Ampacity

The calculation of transmission line ampacity may be simplified if steady-state conditions are assumed. The following assumptions are made in steady-state analysis:

---

* V.T. Morgan, The radial temperature distribution and effective radial thermal conductivity in bare solid and stranded conductors, *IEEE Transactions on Power Delivery*, Volume 5, pp. 1443–1452, July 1990
The thermal behaviour of overhead conductors. Section 3: Mathematical model for evaluation of conductor temperature in the unsteady state, Cigré Working Group WG 22.12 report, *Électra* No. 174, October 1997.

- Conductor temperature remains constant for one hour.
- Conductor current remains constant for one hour.
- Ambient temperature, solar radiation, wind speed, and wind direction are constant for one hour.

The steady-state solution is obtained by setting $\frac{dT_{av}}{dt} = 0$ in (3.5), resulting in the conductor heat balance equation:

$$P_j + P_s + P_m - P_r - P_c = 0 \tag{3.6}$$

By substitution,

$$P_j + P_m = I^2 R_{ac}(Tc) \tag{3.7}$$

The following steady-state solution of conductor ampacity (I) is obtained:

$$I = \sqrt{\frac{P_r + P_c - P_s}{R_{ac}(Tc)}} \tag{3.8}$$

Since the AC resistance of ACSR conductor varies as a function of conductor current (Appendix 3), the ampacity of ACSR conductor is calculated by iteration as shown in Example 1.

Description of symbols:

I = ampacity, A

$$R_{ac} = R_{dc} k_{ac} [1 + \alpha_0 (T_c - T_0)] \tag{3.9}$$

R_{ac} = AC resistance of conductor, ohm/m (R_{ac} may be obtained directly from conductor manufacturer's data sheet or calculated as shown in the Appendix).

R_{dc} = DC resistance of conductor at the reference temperature T_0, ohm/m

$k_{ac} = \dfrac{R_{ac}}{R_{dc}}$

α_0 = temperature coefficient of resistance, /°C
T_c = conductor temperature, °C
T_0 = reference conductor temperature, °C
P_s = heat gains by solar radiation, W/m

$$P_s = \alpha_s D(S_b + S_d) \tag{3.10}$$

α_s = coefficient of solar absorption
D = conductor diameter, m

Theory of Transmission Line Ampacity

S_b = beamed solar radiation, W/m²
S_d = diffused solar radiation, W/m²

$$S_b = S_{ext}\, \tau_b \cos(z) \tag{3.11}$$

$$S_d = S_{ext}\, \tau_d \cos(\theta) \tag{3.12}$$

S_{ext} = 1353 W/m², normal component of the extra terrestrial solar radiation measured outside the earth's atmosphere
τ_b = atmospheric transmittance of beamed radiation
τ_d = atmospheric transmittance of diffused radiation
z = zenith angle, degree
θ = angle of beamed radiation with respect to conductor axis, degree
P_r = heat loss by radiation, W/m

$$P_r = \sigma \varepsilon \pi D\{(T_c + 273)^4 - (T_a + 273)^4\} \tag{3.13}$$

T_a = ambient temperature, °C
σ = 5.67·10⁻⁸, Stephan Boltzman constant, (W/m² K⁴)
ε = Emissivity of conductor
P_c = heat loss by convection, W/m

$$P_c = h \cdot \pi \cdot D(T_c - T_a) \tag{3.14}$$

h = coefficient of heat transfer from conductor surface to ambient air, W/(m²·°C)

$$h = \lambda \cdot Nu \cdot K_{wd}/D \tag{3.15}$$

λ = thermal conductivity of ambient air, W/(m·°C)
Nu = Nusselt number, dimensionless

$$Nu = 0.64\, Re^{0.2} + 0.2\, Re^{0.61} \tag{3.16}$$

Re = Reynolds number, dimensionless

$$Re = D \cdot w_s/v_f \tag{3.17}$$

w_s = wind speed, m/s
v_f = kinematic viscosity of air, m²/s
k_{wd} = wind direction correction factor given by (Davis. 1977)

$$K_{wd} = 1.194 - \sin(\omega) - 0.194 \cos(2\omega) + 0.364 \sin(2\omega) \tag{3.18}$$

ω = wind direction with respect to conductor normal, degree

The AC resistance, R_{ac}, of a conductor is generally available from the manufacturer's catalog for standard conductor sizes at a certain specified conductor temperature. The value of the AC resistance, R_{ac}, of the conductor of any size for any temperature may be calculated by the procedure given in Appendix 3. As shown in the Appendix, the AC resistance of conductor is calculated from the current distribution inside the conductor, conductor construction, and the magnetic properties of the steel core in ACSR.

The AC resistance of conductors having a steel core can be 5 to 15% or more higher than its DC resistance due to the current induced in the steel core as shown in Appendix 3. For stranded conductors without a magnetic core, the AC resistance may be 2 to 5% higher than the DC resistance due to skin effect.

A flow chart of the steady-state current method is shown in Figure 3.1, and a numerical application is given in Example 1. A flow chart of the steady-state temperature calculation method is given in Figure 3.2, and a numerical application is given in Example 2. These methods are used in the LINEAMPS program.

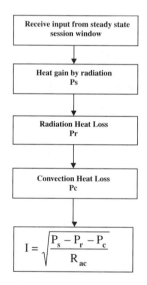

FIGURE 3.1 Flow chart of steady-state ampacity method.

Example 1

Calculate the steady-state ampacity of an ACSR Cardinal conductor. Conductor temperature is 80°C. The meteorological conditions and conductor surface characteristics are as follows:

Ambient temperature = 20°C
Wind speed = 1 m/s
Wind direction = 90° (perpendicular to conductor axis)
Solar radiation = 1000 W/m²
Emissivity = 0.5
Absorptivity = 0.5

Theory of Transmission Line Ampacity

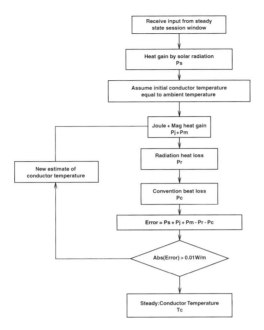

FIGURE 3.2 Flow chart of steady state conductor temperature method.

Solution

Calculate joule heat gain P_j

$$P_j = I^2 \cdot k_{ac} \cdot Rdc_{20}\{1 + \alpha_0(T_c - T_0)\}$$

$$P_j = I^2 \cdot k_{ac} \cdot 0.5973 \cdot 10^{-3} \{1 + 0.004(80 - 20)\}$$

$$P_j = I^2 \cdot k_{ac} \cdot 0.0741 \cdot 10^{-3} \text{ W/m}$$

Calculate solar heat gain P_s

$$P_s = \alpha_s \cdot D \cdot F_s$$

$$P_s = 0.5 \cdot 30.39 \cdot 10^{-3} \cdot 1000$$

$$P_s = 15.19 \text{ W/m}$$

Calculate convection heat loss by wind P_c

$$P_c = k_f \cdot Nu \cdot \pi(T_c - T_a)$$

$$k_f = 2.42 \cdot 10^{-2} + T_f \cdot 7.2 \cdot 10$$

$$T_f = 0.5(80 + 20) = 50°C$$

$$k_f = 2.42 \cdot 10^{-2} + 50 \cdot 7.2 \cdot 10^{-1} = 0.0278 \text{ W/(m·°K)}$$

$$Nu = 0.64 \cdot Re^{0.2} + 0.2 \cdot Re^{0.61}$$

$$Re = \frac{M_{air} \cdot v \cdot D}{vf_h}$$

M_{air} = air density = 1.103 kg/m³

v = 1 m/s

$$vf_h = vf_o \left[1 - H \frac{(6.5 \cdot 10^{-3})}{288.16} \right]^{-5.2561}$$

$vf_o = 1.32 \cdot 10^{-5} + T_f \cdot 9.5 \cdot 10^{-8}$

$vf_o = 1.32 \cdot 10^{-5} + 50 \cdot 9.5 \cdot 10^{-8} = 1.795 \cdot 10^{-5}$ m²/s

H = Altitude, m

Altitude at sea level, H = 0

$vf_h = vf_o = 1.795 \cdot 10^{-5}$ m²/s

$$Re = \frac{1.103 \cdot 30.39 \cdot 10^{-3}}{1.795 \cdot 10^{-5}} = 1867$$

$Nu = 0.64(1867)^{0.2} + 0.2(1867)^{0.061} = 23$

$P_c = 0.0278 \cdot 30.39 \cdot 10^{-3} \cdot \pi \cdot (T_c - T_a) = 0.0278 \cdot 23 \cdot \pi \cdot (80-20)$

$P_c = 120$ W/m

Heat lost by radiation P_r

$P_r = s \cdot \varepsilon \cdot \pi \cdot D[(T_c + 273)^4 - (T_a + 273)^4]$

$P_r = 5.67 \cdot 10^{-8} \cdot 0.5 \cdot \pi \cdot 30.39 \cdot 10^{-3}[(T_c + 273)^4 - (T_a + 273)^4]$

Pr = 22.08 W/m

By substitution in the steady-state heat balance equation,

$$P_j + P_s = P_c + P_r$$

We obtain,

$$I^2 \cdot k_{ac} \cdot 0.0741 + 15.19 = 120 + 2.08$$

$$I = \sqrt{\frac{120 + 22.08 - 15.19}{k_{ac} \cdot 0.0741 \cdot 10^{-3}}}$$

assume $k_{ac} = 1$,

$$I = 1308 \text{ A}$$

For this current $k_{ac} = 1.16$

The revised value I from (3.8) is,

$$I = 1214 \text{ A}$$

For this current $k_{ac} = 1.14$

This process is repeated until convergence, and the final value of current is found to be,

$$I = 1220 \text{ A}$$

The calculation of AC resistance of ACSR Cardinal conductor is shown in the Appendix 1.

Example 2

Calculate the temperature of an ACSR Cardinal conductor. Conductor current is 1220 A, and all other conditions are the same as in Example 1.

Solution

From (3.8) and (3.9) we have,

$$P_j = I^2 \cdot k_{ac} \cdot Rdc_{20} \left\{ 1 + \alpha_0(T_c - T_0) \right\}$$

For I = 1220 A we obtain $k_{ac} = 1.14$, and the joule heat gain P_j is then calculated as,

$$P_j = 1220^2 \cdot 1.14 \cdot 0.05973 \cdot 10^{-3} \left\{ 1 + 0.004(T_c - 20) \right\}$$

From Example 1 we obtain the value of solar heat gain P_s,

$$P_s = 15.19 \text{ W/m}$$

Using the value of k_f and Nu calculated in Example 1, the convection heat loss P_c is obtained from (3.14):

$$P_c = 0.0278 \cdot 23 \cdot \pi \cdot (T_c - 20)$$

The heat loss by radiation P_r is given by (3.13):

$$P_r = 5.67 \cdot 10^{-8} \cdot 0.5 \cdot \pi \cdot 30.39 \cdot 10^{-3} \left[(T_c + 273)^4 - (T_a + 273)^4 \right]$$

The conductor temperature T_c is calculated from the steady-state heat balance equation:

$$P_j + P_s = P_c + P_r$$

By substitution in the above equation we obtain:

$$1220^2 \cdot 1.14 \cdot 0.05973 \cdot 10^{-3} \{1 + 0.004(T_c - 20)\} + 15.19$$
$$= 0.0278 \cdot 23 \cdot \pi (T_c - 20) + 5.67 \cdot 10^{-8} \cdot 0.5 \cdot \pi \cdot 30.39 \cdot 10^{-3}$$
$$[(T_c + 273)^4 - (20 + 273)^4]$$

The above equation is solved for T_c by iteration by giving an initial value $T_c = T_a$. The converged value of T_c is found to be 80°C. A direct solution of steady-state conductor temperature T_c is also obtained from the following quartic equation (Davis, 1977):

$$a_1 \cdot T_c^4 + a_2 \cdot T_c^3 + a_3 \cdot T_c^2 + a_4 \cdot T_c + k_5 = 0 \tag{3.19}$$

where,

$$a_1 = \pi \cdot D \cdot \varepsilon \cdot \sigma$$
$$a_2 = a_1 \cdot 4 \cdot 273$$
$$a_3 = a_1 \cdot 6 \cdot 273^2$$
$$a_4 = a_1 \cdot 4 \cdot 273^3 + \pi \cdot \lambda \cdot Nu$$
$$a_5 = -\left(P_j + P_s + a_1 \cdot T_a^4 + a_2 \cdot T_a^3 + a_3 \cdot T_a^2 + a_4 \cdot T_a\right)$$

and obtain

$$T_c = 80°C$$

The result of conductor temperature T_c obtained by the direct solution of the quartic equation (3.19) is also found to be 80°C.

3.2.4 Dynamic Ampacity

The transmission line conductor is assumed to be in the dynamic state when there is a short-term overload on the line due to line energization or a step change in load. The duration of such overload condition is generally less than 30 minutes. In the dynamic state, the heat storage capacity of the conductor is considered, which allows

Theory of Transmission Line Ampacity

higher than normal static line loading to be allowed on the line for a short duration. The temperature of the conductor in the dynamic state is obtained by the solution of the following differential equation (3.5):

$$M \cdot c_p \frac{dT_{av}}{dt} = P_j + P_s + P_m - P_r - P_c \tag{3.20}$$

The nonlinear differential equation can be solved numerically by Euler's method as follows (Davidson, 1969):

$$T_{av} = \int_0^t \frac{(P_j + P_s + P_m - P_r - P_c)dt}{M \cdot c_p} + T_i \tag{3.21}$$

where,

T_i = initial temperature

By selecting a suitable time interval $\Delta t = dt$, we can replace the above integral by a summation such that,

$$T_{av} = \sum_0^t \frac{(P_j + P_s + P_m - P_r - P_c)\Delta t}{M \cdot c_p} + T_i \tag{3.22}$$

The above equation is solved by the algorithm shown in the flow chart of Figure 3.3a, which is suitable for real-time calculations. A solution of the above nonlinear differential equation by the Runge Kutte method is given by Black (Black et al., 1983).

Direct Solution of Dynamic Conductor Temperature

A direct solution of the nonlinear differential equation is possible by making some simplifying assumptions to linearize the equation. We define an overall heat transfer coefficient h_o (Dalle et al., 1979) such that,

$$P_c + P_r = \pi \cdot D \cdot h_o \cdot (T_c - T_a) \tag{3.23}$$

Combining joule and magnetic heating,

$$P_j + P_m = I^2 \cdot R_{ac} = I^2 \cdot k \cdot Rdc_{20} \{1 + a_0 (T_{av} - T_0)\} \tag{3.24}$$

we obtain,

$$M \cdot c_p \frac{dT_{av}}{dt} = I^2 \cdot k \cdot Rdc_{20}\{1 + \alpha_0(T_{av} - T_0)\} + \alpha_s \cdot D \cdot F_s - \pi \cdot D \cdot h_o \cdot (T_{aw} - T_s) \tag{3.25}$$

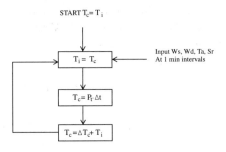

$P_t = P_j + P_m + P_s - P_c - P_r$

$\Delta t = 60s$

T_c = Conductor temperature

$\Delta T_c = T_c + T_i$

T_i = Initial temperature

Figure 3.3.a Real-time calculation dynamic conductor temperature

FIGURE 3.3a Real time calculation of dynamic conductor temperature.

Conductor Temperature

The solution of the linear differential equation (3.25) (Dalle et al., 1979),

$$T_{ch(i)} = \theta_1 - (\theta_1 - T_{ch(i-1)}) \cdot \exp(-\Delta t/\tau_h) \tag{3.26}$$

$$T_{cc(i)} = \theta_2 - (\theta_2 - T_{cc(i-1)}) \cdot \exp(-\Delta t/\tau_h) \tag{3.27}$$

Where,

$T_{ch(i)}$ = conductor temperature during heating
$T_{cc(i)}$ = conductor temperature during cooling

$$\theta_1 = \frac{R_{ac} \cdot I_1^2 (1 - \alpha_0 \cdot T_{ref}) + D \cdot (P_s + \pi \cdot h_o \cdot T_a)}{\pi \cdot D \cdot h_o - \alpha_0 \cdot R_{ac} \cdot I_1^2} \tag{3.28}$$

$$\theta_2 = \frac{R_{ac} \cdot I_2^2 (1 - \alpha_0 T_{ref}) + D \cdot (P_s + \pi \cdot h_o \cdot T_a)}{\pi \cdot D \cdot h_o - \alpha_0 \cdot R_{ac} \cdot I_2^2} \tag{3.29}$$

The heating time constant is given by,

$$\tau_h = \frac{M \cdot c_p}{\pi \cdot D \cdot h_o - \alpha_0 \cdot R_{ac} \cdot I_1^2} \tag{3.30}$$

Theory of Transmission Line Ampacity

The cooling time constant is given by,

$$\tau_c = \frac{M \cdot c_p}{\pi \cdot D \cdot h_0 - \alpha_0 \cdot R_{ac} \cdot I_2^2} \quad (3.31)$$

The coefficient of heat transfer during heating is,

$$h_o = \frac{P_{j1} + P_s}{\pi \cdot D \cdot \Delta T_{c1}} \quad (3.32)$$

The coefficient of heat transfer during cooling is,

$$h_c = \frac{P_{j2} + P_s}{\pi \cdot D \cdot \Delta T_{c2}} \quad (3.33)$$

$$P_{j1} = I_1^2 \cdot R_{ac} \quad (3.34)$$

$$P_{j2} = I_2^2 \cdot R_{ac} \quad (3.35)$$

I_1 = overload current
I_2 = post overload current

$$\Delta T_{c1} = T_{c1} - T_a \quad (3.36)$$

$$\Delta T_{c2} = T_{c2} - T_a \quad (3.37)$$

T_{c1} = steady-state preload conductor temperature
T_{c2} = steady-state overload conductor temperature
Δt = time step, $t_i - t_{i-1}$
A = sectional area of conductor, m^2
D = diameter of conductor, m
α_0 = temperature coefficient of resistance, /°C
R_0 = DC resistance of conductor at reference temperature T_{ref}, ohm/m
T_{ref} = reference temperature, generally 20°C or 25°C
T_a = ambient temperature, °C
R_{ac} = AC resistance of conductor, ohm/m
c_p = specific heat capacity, (J/kg °K) at 20°C. For elevated temperature operation the specific heat may be calculated by, $c_p(Tc) = c_p(T_{20})\{1 + \beta(T_c - T_{20})\}$
β = temperature coefficient of specific heat capacity, /°C (Table 1).

A Flow Chart of Dynamic Temperature method is given in Figure 3.3b and a numerical application is presented in Example 3.

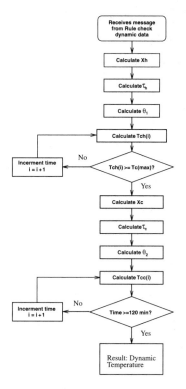

FIGURE 3.3b Flow chart of dynamic temperature method used in the LINEAMPS program.

Direct Solution of Dynamic Ampacity

From Equations (3.27) and (3.28) we may obtain the maximum value of dynamic ampacity approximately as follows:

$$I = \sqrt{\frac{\{T_{max} - T_{initial}\exp(-t/\tau)\}}{C_1\{1-\exp(-t/\tau)\}} - C2} \qquad (3.38)$$

$$C_1 = \frac{R_{ac}(1-\alpha_0 T_{ref})}{\pi \cdot \gamma \cdot D - I^2 \cdot R_{ac} \cdot \alpha_0} \qquad (3.39)$$

$$C_2 = \frac{D(\alpha_s \cdot P_s + \pi \cdot \gamma \cdot T_a)}{R_{ac} \cdot (1-\alpha_0 \cdot T_{ref})} \qquad (3.40)$$

I = dynamic ampacity, A
T_{max} = maximum conductor temperature, °C
$T_{initial}$ = intial conductor temperature, °C

Theory of Transmission Line Ampacity

τ = conductor heating time constant, s
t = duration of overload current, s

A numerical application of dynamic ampacity method is presented in Example 3.4. Formulas for calculation of constants of conductors composed of different material are given in Table 3.1.

TABLE 3.1
(Cigré 1999)

Constant[1]	Formula[2]
Resistivity, ρ, $\Omega \cdot m$	$\rho = \dfrac{\rho_a \rho_s (A_a + A_s)}{\rho_a A_s + \rho_{sd} A_a}$
Temperature coefficient of resistance, $\alpha /° K$	$\alpha = \dfrac{\alpha_a \alpha_s \left(\dfrac{\rho_a}{A_s} + \dfrac{\rho_s}{A_a}\right) + \alpha_a \left(\dfrac{\rho_s}{A_s}\right) + \alpha_s \left(\dfrac{\rho_a}{A_a}\right)}{\dfrac{\rho_a}{A_a} + \dfrac{\rho_s}{A_s} + \alpha_a \left(\dfrac{\rho_a}{A_a}\right) + \alpha_s \left(\dfrac{\rho_s}{A_s}\right)}$
Specific heat, c, J/(kg · °K)	$c = \dfrac{c_a m_a A_a + c_s m_s A_s}{m_a A_a + m_s A_s}$
Temperature coefficient of specific heat, $\beta /° K$	$\beta = \dfrac{c_a m_a \beta_a + c_s m_s \beta_m}{m_a \beta_a + m_s \beta_s}$
Mass, kg/m	$M = \dfrac{A_a m_a + A_s m_s}{A_a + A_s}$

[1]Constants are calculated at 20°C.
[2]Subscripts a, s are for aluminum and steel

m_a, m_s are mass density of aluminum and steel respectively, kg/m³

A = area, m²

Result of Conductor Temperature in Dynamic State

The result of conductor temperature vs. time in the dynamic state is obtained from (3.26) and presented in Figure 3.4. In this example the analysis is carried out by selecting a typical transmission line Zebra conductor by using the Line Ampacity System (LINEAMPS) software developed by the author.*

---

* Anjan K. Deb. Object-oriented expert system estimates transmission line ampacity, *IEEE Computer Application in Power*, Volume 8, Number 3, July 1995.

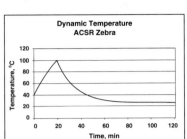

Dynamic Conductor Temperature: ACSR Zebra

FIGURE 3.4 Dynamic conductor temperature as a function of time for ACSR Zebra conductor due to a step change in current.

Example 3.3

Calculate the temperature of ACSR Cardinal conductor when 1475 A overload current is passed through the conductor for 20 minutes. Normal load current is 1260 A. All other conditions are as follows:

Ambient temperature = 20°C
Wind speed = 1 m/s
Wind direction = 90°
Sun = 1000 W/m²
Emissivity = 0.5
Solar absorption = 0.5
Maximum average conductor temperature = 100°C
Steady-state normal conductor temperature at 1260 A = 80°C

Solution

The overall heat transfer coefficient h_o remains fairly constant for a given set of meteorological conditions within a range of conductor temperatures and evaluated by (3.33),

$$h_o = \frac{P_j P_s}{\pi \cdot D \cdot (T_c - T_a)}$$

substituting for P_j and P_s we have,

$$h_o = \frac{I^2 \cdot R_{ac} + \alpha_s \cdot D \cdot F_s}{\pi \cdot D \cdot (T_c - T_a)}$$

$$h_o = \frac{1260^2 \cdot 0.05973 \cdot 10^{-3} \cdot 1.1 \cdot \{1 + 0.004(80-20)\} + 0.5 \cdot 30.39 \cdot 10^{-3} \cdot 1000}{\pi \cdot 30.37 \cdot 10^{-3} \cdot (80-20)}$$

$$h_o = 25.3 \; W/(m^2 \cdot °C)$$

Theory of Transmission Line Ampacity

Substituting values in (3.26),

$$T_{ch(i)} = \theta_1 - \left\{ (\theta_1 - T_{ch(i-1)}) \exp\left(\frac{-t}{\tau_h}\right) \right\}$$

where,

$$\theta_1 = \frac{R_{ac} \cdot I^2 (1 - \alpha_0 \cdot T_{ref}) + D \cdot (P_s + \pi \cdot h_o \cdot T_a)}{\pi \cdot D \cdot h_o - \alpha_0 \cdot R_{ac} \cdot I^2}$$

we obtain,

$$\theta_1 = \frac{0.05973 \cdot 10^{-3} \cdot 1.1 \cdot 1475^2 (1 - 0.004 \cdot 20) + 30.39 \cdot 10^{-3} \cdot (0.5 \cdot 1000 + \pi \cdot 25.3 \cdot 20)}{\pi \cdot 30.39 \cdot 10^{-3} \cdot 25.3 - .004 \cdot 0.05973 \cdot 10^{-3} \cdot 1.1 \cdot 1475^2}$$

$$\theta_1 = 106°C$$

$$\tau_h = \frac{m \cdot c_p}{\pi \cdot D \cdot h_o - \alpha_0 \cdot R_0 \cdot I^2}$$

$$\tau_h = \frac{1.828 \cdot 826}{\pi \cdot 30.39 \cdot 10^{-3} \cdot 25.3 - .004 \cdot 0.05973 \cdot 10^{-3} \cdot 1.1 \cdot 1475^2}$$

$$\tau_h = 822 \text{ s}$$

An average temperature of conductor after 20 minute is obtained from (3.26),

$$T_{av} = 106 - (106 - 80) \cdot \exp\left(\frac{-1200}{822}\right)$$

$$T_{av} = 100°C$$

Example 3.4

Calculate the dynamic ampacity of ACSR Cardinal conductor for 20 minutes. All other conditions are the same as in Example 3.1.

Solution

The dynamic ampacity is calculated directly from (3.38),

$$I = \sqrt{\frac{\{T_{max} - T_{initial} \exp(-t/\tau)\}}{C_1 \{1 - \exp(-t/\tau)\}} - C2}$$

$$C_1 = \frac{R_{ac}(1 - \alpha_0 T_{ref})}{\pi \cdot \gamma \cdot D - I^2 \cdot R_{ac} \cdot \alpha_0}$$

Since current I is not known, we assume a steady-state current I = 1440 A at 100°C to calculate C1,

$$C_1 = \frac{1.1 \cdot 0.0593 \cdot 10^{-3}(1 - 0.004 \cdot 20)}{\pi \cdot 25.2 \cdot 30.39 \cdot 10^{-3} - 1440^2 \cdot 1.1 \cdot 0.0593 \cdot 10^{-3} \cdot 0.004}$$

$$C1 = 3.24310 \cdot 10^{-5}$$

$$C2 = \frac{D \cdot (\alpha_s \cdot P_s + \pi \cdot \lambda \cdot T_a)}{R_{ac} \cdot (1 - \alpha_0 \cdot T_{ref})}$$

$$C2 = \frac{30.39 \cdot 10^{-3} \cdot (0.5 \cdot 1000 + \pi \cdot 25.2 \cdot 20)}{1.1 \cdot 0.0593 \cdot 10^{-3} \cdot (1 - 0.004 \cdot 20)}$$

$$C2 = 1.04810 \cdot 10^5$$

Substituting in (3.38) we obtain the value of dynamic ampacity I,

$$I = \sqrt{\frac{\{100 - 80 \cdot \exp(-1200/822)\}}{3.23 \cdot 10^{-5}\{1 - \exp(-1200/822)\}} - 1.08 \cdot 10^6}$$

$$I = 1484 \text{ A}$$

For greater accuracy we may recalculate C1 with the new value of current I. The final value of dynamic ampacity is found to be 1475 A, which is the same as Example 3.

3.2.5 Transient Ampacity

Transient conditions arise when there is short-circuit or lightning current. The duration of transient current is generally in the range of milliseconds as most power system faults are cleared within few cycles of the 60 Hz frequency. During this time, adiabatic condition is assumed (Cigré, 1999) when there is no heat exchange with the exterior.

Algorithm for the Calculation of Transient Conductor Temperature

Transient conductor temperature response due to short-circuit current is obtained from the solution of the following differential equation:

$$M \cdot c_p \frac{dT_{av}}{dt} = P_j + P_m \qquad (3.41)$$

where,

$$P_j + P_m = I_{sc}^2 R_{ac}\left[1+\alpha_0(T_c - T_0)\right]$$

During adiabatic condition there is no heat exchange with the exterior therefore,

$P_s = 0$
$P_r = 0$
$P_c = 0$

The solution of the differential equation is given by,

$$T_c = T_i + \left(T_0 - \frac{1}{\alpha_0}\right)\left[1 - \exp\left(\frac{\alpha_0 R_{ac} I_{sc}^2 t}{M \cdot c_p}\right)\right] \quad (3.42)$$

Where,

T_i = initial conductor temperature, °C
t = time, s
T_0 = reference temperature, °C
α_0 = temperature coefficient of DC resistance of conductor, /°C
R_{ac} = AC resistance of conductor at reference temperature T_0, ohm/m
I_{sc} = short circuit current, A
M = conductor mass, kg/m
c_p = specific heat of conductor, J/Kg · °K

Equation (3.42) provides the temperature of the conductor during heating by a short circuit current. The temperature during cooling of the conductor is obtained from the dynamic equation.

A flow chart of the transient ampacity method is shown in the Figure 3.5 and a numerical application is shown in Example 5.

Example 5

Calculate the temperature of ACSR Cardinal conductor after a short-circuit current of 50 kA is applied through the conductor for 1 second. The conductor was carrying 1260 A steady-state current when the short-circuit current was applied. All other conditions are the same as in Example 4.

Solution

From Example 4 we obtain the initial temperature of the conductor to be equal to 80°C when the short-circuit current is applied. The following additional data were calculated in Example 4:

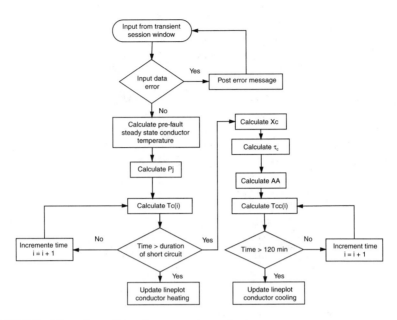

FIGURE 3.5 Flow chart of transient ampacity method.

cp = 826 J/(kg · °C)
R_{ac20} = 6.57 · 10^{-5}

From (3.42) we obtain the temperature of ACSR Cardinal conductor for the following condition:

Short circuit-current, I_{sc} = 50 kA
Duration of short-circuit current, t = 1_s

$$T_c = T_i + \left(T_o - \frac{1}{\alpha_0}\right)\left[1 - \exp\left(\frac{\alpha_0 R_{ac} I_{sc}^2 t}{M \cdot c_p}\right)\right]$$

By substitution of values in the above equation, the temperature of the conductor Tc is calculated as,

$$T_c = 80 + \left(20 - \frac{1}{.004}\right)\left[1 - \exp\left(\frac{0.004 \cdot 6.57 \cdot 10^{-5} \cdot (50 \cdot 10^3)^2 \cdot 1}{1.828 \cdot 826}\right)\right]$$

T_c = 205°C

Theory of Transmission Line Ampacity

The above example shows the importance of clearing faults by a high-speed fault protection system using modern circuit breakers and protective relaying that can detect and clear faults within a few cycles.

Result of Conductor Temperature in Transient State Calculated by Program

Results obtained by the application of the transient ampacity algorithm by using the LINEAMPS program are presented in Figure 3.6. Conductor temperature as a function of time is shown by a line graph when a short-circuit current equal to 50 kA is applied for 0.5s. The conductor is ACSR Zebra.

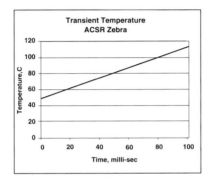

FIGURE 3.6 Transient temperature as a function of time for ACSR Zebra conductor due to a short-circuit current.

3.2.6 RADIAL CONDUCTOR TEMPERATURE

In the previous section, dynamic ampacity calculations were carried out by assuming average conductor temperature. Often, the surface temperature of a conductor is available by measurement, and core temperature is required to calculate sag. The calculation of the radial temperature differential in the conductor is also required for dynamic ampacity calculation. Radial temperature gradient is particularly important for high-ampacity transmission line conductors since they are capable of operating at high temperatures. For high value of ampacity, substantial radial temperature differences from 1 – 5°C were measured in the wind tunnel. Based on the radial temperature differential an average value of conductor temperature can be estimated. In this section the radial temperature of the conductor is derived from the general heat equation. From the general heat equation (3.1) we obtain

$$\frac{\partial^2 T}{r^2} + \frac{1}{r}\frac{\partial T}{\partial r} + \frac{1}{r^2}\cdot\frac{\partial^2 T}{\partial \phi^2} + \frac{\partial^2 T}{\partial z^2} + \frac{q(r)}{k_r} = \frac{1}{\alpha}\cdot\frac{\partial T}{\partial t} \tag{3.43}$$

For a 1m-long cylindrical conductor we may assume that,

$$\frac{\partial T}{\partial \phi} = 0$$

$$\frac{\partial T}{\partial z} = 0$$

In the steady state,

$$\frac{\partial T}{\partial t} = 0$$

Assuming constant heat generation per unit volume, $q(r) = q$ = constant. By the application of above conditions we obtain,

$$\frac{\partial^2 T}{\partial r^2} + \frac{1}{r}\frac{\partial T}{\partial r} + \frac{q(r)}{k_r} = 0 \qquad (3.44)$$

where,

r = radial distance from conductor axis, m
k_r = radial thermal conductivity, W/(m · °K)
q is the internal heat generation by unit volume obtained by,

$$q = \frac{I^2 R_{ac}}{A_{al}} \qquad (3.45)$$

For homogeneous conductors the following boundary conditions are applied,

$$T(r) = T_s \text{ at } r = r_s$$

r_s = conductor radius, m

$$\frac{\partial T(r)}{\partial r} = 0 \text{ at } r = 0$$

The solution to (3.44) is then,

$$T(r) - T_s = r_s^2 I^2 R_{ac}\left\{1 - \left(\frac{r}{r_s}\right)^2\right\} \qquad (3.46)$$

Theory of Transmission Line Ampacity

Substituting,

$$A_{al} = \pi r_s^2$$

$$A_{al} = \text{Aluminum area, m}^2$$

$$P_j + P_m = I^2 R_{ac}$$

The radial temperature difference from conductor core to surface ΔT in homogeneous conductor is obtained as,

$$T(o) - T_s = \Delta T(AAC) = \frac{P_j + P_m}{4\pi k_r} \quad (3.47)$$

For bimetallic conductor (ACSR), the boundary conditions are,

$$T(r) = T_s \text{ at } r = r_s$$

$$\frac{\partial T(r)}{\partial r} = 0 \text{ at } r = r_c$$

By the application of the above boundary conditions for ACSR conductor, the radial conductor differential is obtained by,

$$T_c - T_s = \frac{r_s^2 I^2 R_{ac}}{A_{al} 4 k_r} \left\{ 1 - \left(\frac{r_c}{r_s}\right)^2 + 2\left(\frac{r_c}{r_s}\right) \ln\left(\frac{r_c}{r_s}\right) \right\} \quad (3.48)$$

where,

T_c = conductor core temperature, °C
T_s = conductor surface temperature, °C

3.3 CHAPTER SUMMARY

Starting with a three-dimensional transmission line conductor thermal model, a differential equation of conductor temperature with respect to time is developed in this chapter. Steady-state solutions of the differential equation are given for the calculation of conductor ampacity and conductor temperature. Differential equations are developed for dynamic and transient conditions, and their closed form solutions are given. The radial temperature differential in the conductor due to the difference in the surface and core temperature is also derived. Algorithms for the calculation of transmission line conductor ampacity and temperature are presented with worked-out practical examples. The AC resistance of ACSR conductors increases with conductor temperature as well as conductor current. The calculation of AC resistance of ACSR conductor, including magnetic heating and current redistribution in the different layers of the conductor, is presented in Appendix 3 at the end of this chapter with a numerical application.

Appendix 1
AC Resistance of ACSR

The AC resistance of conductors having magnetic cores is greater than their DC resistance because of the transformer action created by the spiraling effect of current in the different layers of aluminum wires. The increase in the AC resistance of ACSR conductors are mainly due to current redistribution in the aluminum wire layers, and the magnetic power loss in the steel core due to eddy current and hysteresis loss. Therefore, the AC resistance of ACSR conductors may be considered to be composed of the following:

1. DC resistance
2. Increment in resistance due to current redistribution
3. Increment in resistance due to magnetic power losses in the steel core

The resistance and inductance model of a three-layer ACSR conductor is shown in Figure A1.1 (Vincent, M., 1991), (Barrett et al., 1986). As shown in the figure, the reactance of each layer of aluminum wire is due to the self-inductance, L_{nn}; mutual inductance; $L_{m,n}$, due to the longitudinal flux; and the circular inductance, L_c, due to the circular flux. The circular inductance model assumes that there is 21% contribution due to inner flux, and 79% contribution by the outer flux of each wire in a layer (Vincent, M., 1991), (Barrett et al., 1986). The longitudinal inductances lead to the longitudinal self reactances X_{mm}, and mutual reactances, $X_{m,n}$. Similarly, the inner and outer circular inductances lead to the inner and outer circular reactances, $Xc_{n,i}$ and $Xc_{n,o}$, respectively.

The current redistribution in the different layers of the aluminum wires are due to longitudinal and circular flux, which are calculated as follows (Vincent, M., 1991), (Barrett et al., 1986).

Longitudinal Flux

The magnetic field intensity, H (A/m), of a wire carrying current, I, is given by Ampere's current law,

$$\oint_c H \cdot dl = N \cdot I \qquad (A\ 1.1)$$

N = number of turns of aluminum wires over the steel core given by,

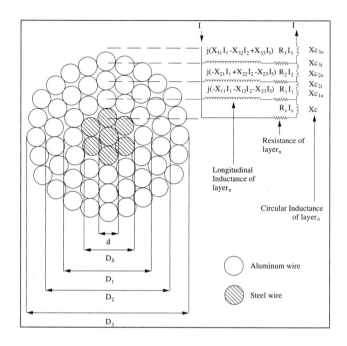

FIGURE A1.1 Electric circuit model of ACSR.

$$N = \frac{1}{s_i} \qquad (A\ 1.2)$$

s_i = lay length of layer i, m

The magnetic flux ϕ due to the magnetic field B (T) is obtained as,

$$\phi = \int_s B \cdot ds \qquad (A\ 1.3)$$

Therefore, for layer n, the magnetic flux, ϕ_n, is,

$$\phi_n = B_n \cdot A_n \qquad (A\ 1.4)$$

B_n = magnetic field, Tesla

Applying,

$$B = \mu H$$

$\mu = \mu_0 \mu_r$

Appendix 1 AC Resistance of ACSR

The flux in layer n is obtained by,

$$\phi_n = \mu_0 \mu_r H_n A_n \tag{A 1.5}$$

$$\phi_n = \mu_0 \left[\left(\pi r_n^2 - A_c \right) + \mu_r A_c \right] \frac{I_n}{S_n} \tag{A 1.6}$$

The self-inductance of layer n, L_{nn} ($\mu_{r(al)} = 1$) is given by,
The self-reactance of layer n, X_{nn} is,

$$X_{nn} = \frac{2\pi f \mu_0 \left[\left(\pi r_n^2 - A_c \right) + \mu_r A_c \right]}{S_n^2} \tag{A 1.7}$$

The mutual inductance of layer m, n is,

$$M_{m,n} = M_{n,m} = \frac{\mu_0 \left[\left(\pi r_n^2 - A_c \right) + \mu_r A_c \right]}{S_n \cdot S_q} \tag{A 1.8}$$

And the mutual reactance of layer m, n is,

$$X_{m,n} = X_{n,m} = \frac{2\pi f \mu_0 \left[\left(\pi r_n^2 - A_c \right) + \mu_r A_c \right]}{S_n \cdot S_q} \tag{A 1.9}$$

Circular Flux

By assuming that the layers currents are concentrated at the center of each layer, the outer circular flux due to layer, n, of length, l, is obtained by,

$$\phi_{n,outer} = \int_0^l \int_{D_n}^{D_{n-1}} \frac{\sum_{n=0}^{N} I_n}{2\pi r} \mu_r \mu_0 \, dr.dz \tag{A 1.10}$$

Which has for solution,

$$\phi_{n,outer} = \frac{\sum_{n=0}^{N} I_n}{2\pi} \mu_r \mu_0 \ln\left(\frac{D_n}{D_n - d} \right) \tag{A 1.11}$$

Similarly the inner circular flux due to layer (n + 1) is obtained as,

$$\phi_{n+1,\text{inner}} = \frac{\sum_{n=0}^{N} I_n}{2\pi} \mu_r \mu_0 \ln\left(\frac{D_{n+1} - d}{D_n}\right) \quad \text{(A 1.12)}$$

It was previously shown (Vincent, M., 1991), (Barrett et al., 1986) that the layer current contributes 21% to the inner flux, and 79% to the outer flux because of current distribution in a wire as shown in Figure A1.1.

The voltage drop per meter along each layer is given by,

The voltage drop V_1 in layer 1 is,

$$\overline{V}_1 = \overline{I}_1 R_1 + j\overline{I}_1 X_{11} - j\overline{I}_2 X_{12} + j\overline{I}_3 X_{13} + j\mu_0 f\left(\overline{I}_s + 0.79\overline{I}_1\right) \ln\left(\frac{D_1}{D_1 - d}\right)$$

$$+ j\mu_0 f\left(\overline{I}_s + \overline{I}_1 + 0.21\overline{I}_2\right) \ln\left(\frac{D_2 - d}{D_1}\right) + j\mu_0 f\left(\overline{I}_s + \overline{I}_1 + 0.79\overline{I}_2\right) \ln\left(\frac{D_2}{D_2 - d}\right)$$

$$+ j\mu_0 f\left(\overline{I}_s + \overline{I}_1 + \overline{I}_2 + 0.21\overline{I}_3\right) \ln\left(\frac{D_3 - d}{D_2}\right) + j\mu_0 f\left(\overline{I}_s + \overline{I}_1 + \overline{I}_2 + 0.79\overline{I}_3\right) \ln\left(\frac{D_3}{D_3 - d}\right)$$

(A 1.13)

The voltage drop V_2 in layer 2 is,

$$\overline{V}_2 = \overline{I}_2 R_2 - j\overline{I}_1 X_{21} - j\overline{I}_2 X_{22} - j\overline{I}_3 X_{23} + j\mu_0 f\left(\overline{I}_s + \overline{I}_1 + 0.79\overline{I}_2\right) \ln\left(\frac{D_2}{D_2 - d}\right)$$

$$+ j\mu_0 f\left(\overline{I}_s + \overline{I}_1 + \overline{I}_2 + 0.21\overline{I}_3\right) \ln\left(\frac{D_3 - d}{D_1}\right) + j\mu_0 f\left(\overline{I}_s + \overline{I}_1 + 0.79\overline{I}_2\right) \ln\left(\frac{D_2}{D_2 - d}\right)$$

$$+ j\mu_0 f\left(\overline{I}_s + \overline{I}_1 + \overline{I}_2 + 0.21\overline{I}_3\right) \ln\left(\frac{D_3 - d}{D_2}\right) + j\mu_0 f\left(\overline{I}_s + \overline{I}_1 + \overline{I}_2 + 0.79\overline{I}_3\right) \ln\left(\frac{D_3}{D_3 - d}\right)$$

(A1.14)

The voltage drop $\overline{V}_3$ in layer 3 is,

$$\overline{V}_3 = \overline{I}_3 R_3 + j\overline{I}_1 X_{31} - j\overline{I}_2 X_{32} + j\overline{I}_3 X_{23} + j\mu_0 f\left(\overline{I}_s + \overline{I}_1 + 0.79\overline{I}_3\right) \ln\left(\frac{D_3}{D_3 - d}\right) \quad \text{(A 1.15)}$$

Appendix 1 AC Resistance of ACSR

The voltage drop $\overline{V}_s$ in the steel core is,

$$\overline{V}_s = \overline{I}_s R_s + j\mu_0 f(\overline{I}_s + 0.21\overline{I}_1)\ln\left(\frac{D_1 - d}{D_c}\right) + j\mu_0 f(\overline{I}_s + 0.79\overline{I}_1)\ln\left(\frac{D_2}{D_1 - d}\right)$$

$$+ j\mu_0 f(\overline{I}_s + \overline{I}_1 + 0.21\overline{I}_2)\ln\left(\frac{D_2 - d}{D_1}\right) + j\mu_0 f(\overline{I}_s + \overline{I}_1 + 0.79\overline{I}_2)\ln\left(\frac{D_2}{D_2 - d}\right)$$

$$+ j\mu_0 f(\overline{I}_s + \overline{I}_1 + \overline{I}_2 + 0.21\overline{I}_3)\ln\left(\frac{D_3 - d}{D_2}\right) + j\mu_0 f(\overline{I}_s + \overline{I}_1 + \overline{I}_2 + 0.79\overline{I}_3)\ln\left(\frac{D_3}{D_3 - d}\right)$$

(A 1.16)

Equations (A1.13)–(A 1.16) may be set up as a set of four simultaneous equations with four unknown currents, $\overline{I}_1, \overline{I}_2, \overline{I}_3, \overline{I}_2$, which satisfy the following conditions:
The sum of layer currents must equal total current $\overline{I}$,

$$\overline{I}_1 + \overline{I}_2 + \overline{I}_3 + \overline{I}_s = \overline{I} \qquad (A\ 1.17)$$

The voltage drop of each layer are equal,

$$\overline{V}_1 = \overline{V}_2 = \overline{V}_3 = \overline{V}_s \qquad (A\ 1.18)$$

From the calculated layer currents we obtain the voltage drop, V. The AC resistance of the conductor is then found by,

$$R_{ac} = \mathrm{Re}\left(\frac{\overline{V}}{\overline{I}}\right)$$

The calculation of the AC resistance of a conductor is carried out iteratively because the complex relative permeability, μ_r, of steel core is a nonlinear function of the magnetic field intensity, H. The magnetic field intensity, H, is a function conductor current. The following relation may be used to calculate complex relative permeability, μ_r, for H ≤ 1000 A/m.

$$\mu_r = [40 - 0.0243|H| + 0.000137|H|^2] - j[5 + 1.03 \cdot 10^{-10}|H|^2]$$

Example 6

Calculate the AC resistance of a 54/7 ACSR Cardinal conductor for the following operating conditions:

Conductor current = 1000 A
Average conductor temperature = 80°C

Solution

Cardinal conductor data:

Number of steel wires, $n_s = 7$
Number of aluminum wires, $n_{al} = 54$
Number of aluminum wires in layer 1 = 12
Number of aluminum wires in layer 2 = 18
Number of aluminum wires in layer 3 = 24
Wire diameter, d = 3.376 mm
Conductor diameter, D = 30.38 mm
Aluminum resistivity, $\rho_a = 0.028126\ \Omega \cdot mm^2/m$
Steel resistivity, $\rho_s = 0.1775\ \Omega \cdot mm^2/m$

The following layer lengths are assumed:

$\lambda_s = 0.253$ m
$\lambda_1 = 0.219$ m
$\lambda_2 = 0.236$ m
$\lambda_3 = 0.456$ m

The dc resistance of Layer i is given by,

$$Rdc_i = \frac{\rho_i \sqrt{1 + \frac{\{\pi(D_i - d)10^{-3}\}}{s_i}}}{A_i n_i} \quad i = 1, 2, 3 \text{ layers}$$

Initial layer current,

$$I_1 = \frac{1}{\left[1 + Rdc_1 \left\{\frac{1}{Rc} + \frac{1}{Rdc_2} + \frac{1}{Rdc_3}\right\}\right]}$$

$$I_2 = \frac{I_1 \cdot Rdc_1}{Rdc_2}$$

$$I_3 = \frac{I_1 \cdot Rdc_1}{Rdc_3}$$

$$I_s = \frac{I_1 \cdot Rdc_1}{R_s}$$

Appendix 1 AC Resistance of ACSR

We obtain initial layer currents,

$I_1 = 327$ A
$I_2 = 489$ A
$I_3 = 652$ A
$I_c = 31$ A

The voltage drop in Layer 1 is,

$$\bar{V}_1 = \bar{I}_1 R_1 + j\bar{I}_1 X_{11} - j\bar{I}_2 X_{12} + j\bar{I}_3 X_{13} + j\mu_0 f\left[\left(\bar{I}_s + 0.79\bar{I}_1\right)\ln\left(\frac{5}{4}\right)\right.$$

$$+ \left(\bar{I}_s + \bar{I}_1 + 0.21\bar{I}_2\right)\ln\left(\frac{6}{5}\right) + \left(\bar{I}_s + \bar{I}_1 + 0.79\bar{I}_2\right)\ln\left(\frac{7}{6}\right)$$

$$+ \left(\bar{I}_s + \bar{I}_1 + \bar{I}_2 + 0.21\bar{I}_3\right)\ln\left(\frac{8}{7}\right) + \left.\left(\bar{I}_s + \bar{I}_1 + \bar{I}_2 + 0.79\bar{I}_3\right)\ln\left(\frac{9}{8}\right)\right]$$

The voltage drop, V_2, in Layer 2 is,

$$\bar{V}_2 = \bar{I}_2 R_2 - j\bar{I}_1 X_{21} - j\bar{I}_2 X_{22} - j\bar{I}_3 X_{23} + j\mu_0 f\left[\left(\bar{I}_s + \bar{I}_1 + 0.79\bar{I}_2\right)\ln\left(\frac{7}{6}\right)\right.$$

$$+ \left(\bar{I}_s + \bar{I}_1 + \bar{I}_2 + 0.21\bar{I}_3\right)\ln\left(\frac{8}{5}\right) + \left(\bar{I}_s + \bar{I}_1 + 0.79\bar{I}_2\right)\ln\left(\frac{7}{6}\right)$$

$$+ \left(\bar{I}_s + \bar{I}_1 + \bar{I}_2 + 0.21\bar{I}_3\right)\ln\left(\frac{8}{7}\right) + \left.\left(\bar{I}_s + \bar{I}_1 + \bar{I}_2 + 0.79\bar{I}_3\right)\ln\left(\frac{9}{8}\right)\right]$$

The voltage drop, $\bar{V}_3$, in Layer 3 is,

$$\bar{V}_3 = \bar{I}_3 R_3 + j\bar{I}_1 X_{31} - j\bar{I}_2 X_{32} + j\bar{I}_3 X_{23} + j\mu_0 f\left(\bar{I}_s + \bar{I}_1 + \bar{I}_2 + 0.79\bar{I}_3\right)\ln\left(\frac{9}{8}\right)$$

The voltage drop, $\bar{V}_s$, in the steel core is,

$$\bar{V}_s = \bar{I}_s R_s + j\mu_0 f\left[\left(\bar{I}_s + 0.21\bar{I}_1\right)\ln\left(\frac{4}{3}\right) + \left(\bar{I}_s + 0.79\bar{I}_1\right)\ln\left(\frac{5}{4}\right)\right.$$

$$+ \left(\bar{I}_s + \bar{I}_1 + 0.21\bar{I}_2\right)\ln\left(\frac{6}{5}\right) + \left(\bar{I}_s + \bar{I}_1 + 0.79\bar{I}_2\right)\ln\left(\frac{7}{6}\right)$$

$$+ \left(\bar{I}_s + \bar{I}_1 + \bar{I}_2 + 0.21\bar{I}_3\right)\ln\left(\frac{8}{7}\right) + \left.\left(\bar{I}_s + \bar{I}_1 + \bar{I}_2 + 0.79\bar{I}_3\right)\ln\left(\frac{9}{8}\right)\right]$$

Where,

$$X_{11} = 2\pi \cdot 60 \cdot \mu_0 \frac{A_c}{s_1^2}$$

$$X_{12} = 2\pi \cdot 60 \cdot \mu_0 \frac{A_c}{s_1 \cdot s_2}$$

$$X_{13} = 2\pi \cdot 60 \cdot \mu_0 \frac{A_c}{s_1 \cdot s_3}$$

$$X_{22} = 2\pi \cdot 60 \cdot \mu_0 \frac{A_c}{s_2^2}$$

$$X_{23} = 2\pi \cdot 60 \cdot \mu_0 \frac{A_c}{s_2 \cdot s_3}$$

$$X_{33} = 2\pi \cdot 60 \cdot \mu_0 \frac{A_c}{s_3^2}$$

The magnetic field, H (A/m), is obtained by,

$$|H| = \frac{|I_1|}{s_1} - \frac{|I_2|}{s_2} + \frac{|I_3|}{s_3}$$

The complex relative permeability, $\overline{\mu}_r$, of the steel core is given by,

$$\overline{\mu}_r = \left[40 - 0.0243|H| + 0.000137|H|^2\right] - j\left[5 + 1.03 \cdot 10^{-10}|H|^4\right]$$

The sum of all layer currents is equal to total current,

$$\overline{I}_1 + \overline{I}_2 + \overline{I}_3 + \overline{I}_s = \overline{I}$$

Voltage drop in each layer is equal,

$$\overline{V}_1 = \overline{V}_2 = \overline{V}_3 = \overline{V}_s$$

The above problem was solved by Mathcad®* Solver giving initial values and the following results were obtained,

$$\overline{V}_1 = \overline{V}_2 = \overline{V}_3 = \overline{V}_s = 0.109 + j0.024$$

---

* Mathcad 8® is registered trademark of Mathsoft, Inc., http://www.mathsoft.com/

Appendix 1 AC Resistance of ACSR

$$\bar{I}_1 = 273 - j104$$

$$\bar{I}_2 = 534 + j62$$

$$\bar{I}_3 = 659 + j52$$

$$\bar{\mu}_r = 131 - j76$$

Current density:

Layer 1 = 2.72 A/m²
Layer 2 = 3.34 A/m²
Layer 3 = 3.08 A/m²

The ac resistance of the conductor is given by,

$$R_{ac} = \text{Re}\left(\frac{\bar{V}}{\bar{I}}\right) = 7.432 \cdot 10^{-5}$$

$$\frac{R_{ac}}{R_{dc}} = \frac{7.432 \cdot 10^{-5}}{6.432 \cdot 10^{-5}} = 1.127$$

The ac/dc ratio, k, is composed of a factor k_1 due to current redistribution in the layers, and a factor k_2 due to magnetic power loss in a ferromagnetic core, and is given by,

$$\frac{R_{ac}}{R_{dc}} = k = k_1 \cdot k_2$$

The current redistribution factor, k_1, is obtained by,

$$k_1 = \frac{|\bar{I}_c|^2 \cdot R_c + |\bar{I}_1|^2 \cdot R_1 + |\bar{I}_2|^2 \cdot R_2 + |\bar{I}_3|^2 \cdot R_3}{I^2 \cdot R_{dc}}$$

The magnetic power loss factor, k_2, is obtained by,

$$k_2 = \frac{k}{k_1}$$

4 Experimental Verification of Transmission Line Ampacity

I have always endeavoured to make experiment the test and controller of theory and opinion.

Michael Faraday on Electricity

4.1 INTRODUCTION

The object of this chapter is to present experimental data on transmission line ampacity for the development and validation of theory, hypotheses, and assumptions. The data presented in this chapter is compiled from the different tests that I have either conducted myself, or were conducted by other people in different research laboratories. As much as possible, the data presented here are from published literature. I have selected Michael Faraday's (1834) quotation for this discussion not only for its general applicability to all experimental research, but also for his particular interest in the subject of electricity and the heating of wires by electric current.

4.2 WIND TUNNEL EXPERIMENTS*

Experiments were carried out at in a wind tunnel to verify conductor thermal modeling for static and dynamic thermal ratings, and to determine the radial thermal conductivity of conductors. Atmospheric conditions of wind speed and ambient temperature were simulated in a wind tunnel that was specially built for these studies. A transmission line conductor was installed in the wind tunnel, and current was passed through it to study the effects of environmental variables on conductor heating.

In Table 4.1, wind tunnel data is compared to the values calculated by the program, showing excellent agreement between measured and calculated values. The measured value of steady-state ampacity is 1213 Amperes in the wind tunnel with 2.4 m/s wind, 90° wind direction, and 43°C ambient temperature. The same

---

* "Wind Tunnel Studies of Transmission Line Conductor Temperatures," by J.F. Hall, Pacific Gas & Electric Co., and Anjan. K. Deb, Consultant, Innova Corporation, presented and published in *IEEE Transactions in Power Delivery*, Vol. 3, No. 2, April 1988, pages 801–812.

value is calculated by the LINEAMPS program. When wind speed is zero, the measured value of ampacity is 687 Amperes compared to 718 Amperes calculated by the program. Due to the inherent uncertainties in the measurement of atmospheric variables, we may easily expect 5 to 10% measurement error. The difference between measurement and calculations are within this range in all of the data presented in Table 4.1.

In addition to the comparison of wind tunnel data with the program, data from several other sources are presented to show their excellent agreement with results obtained by the LINEAMPS program. The data presented in Table 4.1 represents a diverse sampling of line ampacity results obtained in the different regions of the world, including the northern and southern hemispheres of the globe, which have different national standards. For example, Southwire is a well-known conductor manufacturing company in the U.S. PG&E is the largest investor-owned electric utility in the U.S., EDF is the national electric supply company of France, and Dr. Vincent Morgan is a leading authority on conductor thermal rating in Australia (Morgan, 1991). In all of the above examples, the results obtained from the LINEAMPS program compared well with the data presented in Table 4.1.

A sketch of the wind tunnel is given in Figure 4.1, showing the placement of the conductor inside the wind tunnel to achieve different wind angles. A 25-hp motor was used to power four 36-inch fans at 0 to 960 rpm. Wind speed in the range of 0 to 20 mph was generated inside the wind tunnel and measured by propeller-type anemometers manufactured by R. M. Young Co. Two sizes of four-blade polystyrene propellers were used for low and high wind-speed measurements. The smaller propellers were 18 cm in diameter and 30 cm in pitch, and the larger propellers were 23 cm in diameter and 30 cm in pitch.

From wind tunnel experimental data of conductor temperature at various wind speeds, the following empirical relationship between the Nusselt number (Nu) and the Reynolds number (Re) is determined for the calculation of forced convection cooling in conductor:

$$\text{Nu} = \exp\{3.96 - 0.819 \cdot \ell n \, \text{Re} + 0.091 \cdot (\ell n \, \text{Re})^2\} \quad (4.1)$$

$$1000 \leq \text{Re} \leq 15000$$

The results from the above equation are compared to data presented by other researchers in Figure 4.2 with excellent agreement.

When higher-than-normal transmission line ampacity is allowed through a line, it is also necessary to evaluate the probability distribution of conductor temperature. The loss of conductor tensile strength (Mizuno et al., 1998), the permanent elongation of the conductor due to creep (Cigré, 1978), the safety factor of the line, and transmission line sag as a function of the life of the line (Hall and Deb, 1988b) are calculated from the probability distribution of conductor temperature and discussed further in Chapter 5.

Experimental Verification of Transmission Line Ampacity

TABLE 4.1
Ampacity Test Results

Source	Conductor	Sun	Ta °C	Ws m/s	Wd°	Tc °C	Rating Type	Source Amp	LINEAMPS Amp
Southwire	Drake	N	40	0		50	Summer Steady	320	319
Southwire	Drake	Y	40	1.2	90	75	Summer Steady	880	880
Southwire	Drake	Y	40	0.61	90	100	Summer Emergency 15 min	1160	1160
PG&E	Cardinal	Y	43	0.61	90	80	Summer Steady	838	830
Wind Tunnel	Cardinal	N	43	0		78.1	Measured Steady	687	718
Wind Tunnel	Cardinal	N	43	2.4	90	75.8	Measured Steady	1213	1213
EDF	Aster 570	Y	30	1	90	60	Summer Steady	830	826
EDF	Aster 851	Y	15	1	90	60	Winter Steady	1350	1366
EDF	Aster 570	Y	15	1	90	75	Winter Dynamic 20 min	1393	1390
EDF	Aster 570	Y	15	1	90	150	Transient 1 sec	44.5 kA	44.5 kA
Morgan V. T.	Curlew	N	0	0.6	90	80	Winter Steady Night	1338	1324
Morgan V. T.	Curlew	Y	5	0.6	90	80	Winter Steady Noon	1182	1233

Notes:
Ta = Ambient temperature, Degree Celsius
Ws = Wind speed, meter per second
Wd = Wind direction, Degree
Tc = Conductor surface temperature, Degree Celsius
Y = Yes, N = No

Source = Name of company or research publication from where data was obtained for this test.
- Southwire is a trademark of Southwire Company, Carrolton, GA.
- PG&E is the Pacific Gas & Electric Company, San Francisco.
- EDF is Electricité de France, Paris.
- Vincent T. Morgan is author of *Thermal Behavior of Electrical Conductors*, published by Wiley, Inc., New York, 1991.
- LINEAMPS is *Line Ampacity System*, an object-oriented expert line ampacity system. U.S. Patent 5,933,355 issued August 1999 to Anjan K. Deb.
- Wind Tunnel data from: "Wind Tunnel Studies of Transmission Line Conductor Temperatures," *IEEE Transactions on Power Delivery*, Vol. 3, No. 2, April 1988, Authors: J. F. Hall, Anjan K. Deb, J. Savoullis.

4.3 EXPERIMENT IN OUTDOOR TEST SPAN

An outdoor test span is useful for the verification of transmission line sag and tension calculated by the LINEAMPS program. The computer program for the calculation of sag and tension uses the following transmission line conductor change of state equation:

$$\frac{\sigma_2}{E} - \frac{(\varpi \cdot L)^2}{24\sigma_2^2} + \alpha(Tc_2 - Tc_1) + \Delta Ec = \frac{\sigma_1}{E} - \frac{(\varpi \cdot L)^2}{24\sigma_1^2} \quad (4.2)$$

64 Powerline Ampacity System: Theory, Modeling, and Applications

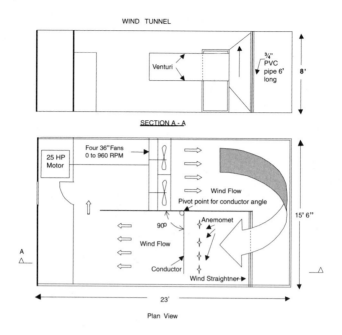

FIGURE 4.1 Wind tunnel.

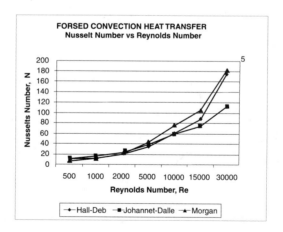

FIGURE 4.2 Forced convection Nusselt number vs Reynolds number relationship obtained from wind tunnel experiments and comparison with results from other researchers.

where,

σ_1, σ_2 = stress at state1 and state2 respectively, kg/mm²
Tc_1, Tc_2 = conductor temperature at state1 and state 2, °C
E = Young's modulus of elasticity, kg/mm²

Experimental Verification of Transmission Line Ampacity

ϖ = specific weight of conductor, kg/m/mm^2
L = span length, m
ΔEc = inelastic elongation (creep) mm/mm
α = coefficient of linear expansion of conductor, °C^{-1}

Results obtained by the application of above equation are presented in Table 4.2.
The sag and tension program is further verified by comparison with field data from various electric power companies* with excellent agreement (Wook, Choi, and Deb, 1997).

TABLE 4.2
Verification of Transmission Line Security
(ACSR Cardinal Conductor)

Tc °C	Wind, Pa	LOS %RTS	Creep µstrain	Tension kN	Safety Factor	Sag m	Life Year
15	1480	4	1200	73.03	2.00	2.85	50
80	–	4	1200	20.56	7.00	10.12	50
100	–	4	1200	19.40	7.47	10.72	50

1 Pascal (Pa) = 0.02 lbf/ft2
1 kN (Kilo Newton) = 224.8 lbf
mstrain = micro strain = mm/km
RTS = Rated Tensile Strength
Tc = Average conductor temperature, °C
Wind = Wind pressure on projected area of conductor, Pa
LOS = Loss of Strength
SF = Safety Factor of conductor
Initial sag after stringing = 9.1m @ 100 °C

As shown in Figure 4.3, the transmission line conductor was energized by a 100 kVA transformer, and the temperature of the conductor was controlled by varying the current passing through it. Conductor sag at midspan was measured by a scale which compared well with the sag calculated by the program.

The loss of strength of aluminum alloy wires was determined experimentally by heating individual wires at elevated temperatures. The results of this experiment are presented in Table 4.3. These experiments were conducted at the EDF laboratory at Paris, France (Deb, 1978).

---

* Thanks are due to Mr. Wally Sun, Transmission Line Engineer, PG&E, San Francisco, CA, for providing transmission line sag and tension data. A report was submitted to Mr. Sun which shows the result of this comparison, June 1990.

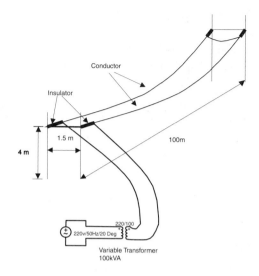

FIGURE 4.3 Test setup of high temperature conductor sag measurement in outdoor test span.

TABLE 4.3
Loss of Strength of Aluminum Alloy Wires at Elevated Temperatures

Conductor Temperature, °C	Duration, hr	Loss of Strength, %
150	100	32.35
150	20	24.71
150	5	7.65
150	4	4.80
150	2	2.40
130	10	2.06
130	10	1.47

Note: Aluminum alloy wire size is 3.45 mm in diameter.

4.4 COMPARISON OF LINEAMPS WITH IEEE AND CIGRÉ

4.4.1 STEADY-STATE AMPACITY

The IEEE* recommends a standard method for the calculation of current-carrying capacity of overhead line conductors based on theoretical and experimental research

---

* IEEE Standard 738-1993. *IEEE Standard for calculating the current-temperature relationship of bare overhead conductors.*

Experimental Verification of Transmission Line Ampacity

carried out by several researchers. Similarly, Cigré* (Conférence International de la Grande Réseaux Electrique), the international conference on large electrical networks, has proposed a method for calculating the thermal rating of overhead conductors. The two methods of ampacity calculation were compared by PG&E engineer N.P. Schmidt,** and the results of this comparison were presented in a 1997 IEEE paper (Schmidt. 1997). The comparisons are based on steady-state conditions only. The study shows that there may be up to 10% variation in the two methods of ampacity calculation.

In this section, LINEAMPS results are compared to the results given by Schmidt (1997) in Figures 4.4–4.9. These results show that the values calculated by LINEAMPS are within 10% of those of the IEEE (IEEE Std. 738, 1993) and Cigré (1997, 1992). The assumptions regarding the transmission line and the various meteorological conditions are presented in Table 4.4 from the IEEE paper.

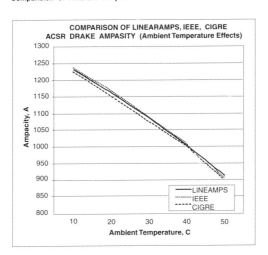

FIGURE 4.4 Ampacity calculated by LINEAMPS program is compared to the IEEE Standard and Cigré method of calculating conductor thermal rating in the steady state. Figure shows the variation of conductor ampacity as a function of ambient temperature. All other assumptions are specified in Table 4.4.

It is appropriate to mention here that these comparisons were made on the assumptions that the meteorological conditions comprised of wind speed, wind direction, sky condition, and ambient temperature are the same all along the transmission line route. It is important to note that IEEE and Cigré provide methods to calculate line ampacity when the ambient conditions are given. They do not include

---

* The thermal behaviour of overhead conductors. Section 1, 2, and 3. Report prepared by Cigré Working Group 22.12. Section 1 and 2, *Electra*, October 1992, Section 3, *Electra*, October 1997.
** N. P. Schmidt. *Comparison between IEEE and Cigré ampacity standards.* IEEE Power Engineering Society conference paper # PE-749-PWRD-0-06-1997. Anjan K. Deb, Discussion contribution to this paper, October 1997.

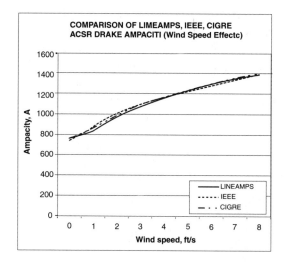

FIGURE 4.5 Ampacity calculated by LINEAMPS program is compared to the IEEE Standard and Cigré method of calculating conductor thermal rating in the steady state. Figure shows the variation of conductor ampacity as a function of wind speed. All other assumptions are specified in Table 4.4.

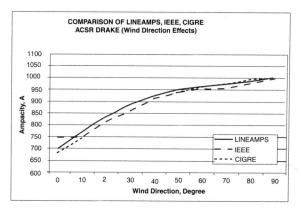

FIGURE 4.6 Ampacity calculated by LINEAMPS program is compared to the IEEE Standard and Cigré method of calculating conductor thermal rating in the steady state. Figure shows the variation of conductor ampacity as a function of wind direction. All other assumptions are specified in Table 4.4.

methods for modeling variations in meteorological conditions along the transmission line route. The different meteorological conditions along the transmission line route are considered in a unique manner by the LINEAMPS program, as stated in a discussion contribution recently prepared by this author (Deb, 1998). Faraday also

Experimental Verification of Transmission Line Ampacity

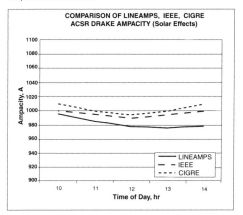

FIGURE 4.7 Ampacity calculated by LINEAMPS program is compared to the IEEE Standard and Cigré method of calculating conductor thermal rating in the steady state. The effect of solar radiation on conductor ampacity is shown as a function of time of day. All other assumptions are specified in Table 4. LINEAMPS considers both direct beam and the diffused solar radiation hence the predicted ampacity is slightly lower than IEEE and Cigré. Diffused radiation was neglected in the comparison made in the IEEE paper. However, the effect of solar radiation on line ampacity is comparatively small when compared to the effects of ambient temperature and wind.

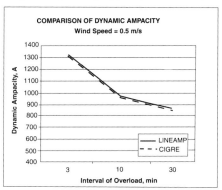

FIGURE 4.8 Ampacity calculated by LINEAMPS program in dynamic state is compared to Cigré method of calculation of dynamic ampacity when wind speed is 0.5 m/s.

realized the problem of changing cooling effects on wire when different parts of the wire are exposed to different cooling conditions when he stated:*

---

* Michael Faraday on Electricity. "On the absolute quantity of electricity associated with the particles or atoms of matter." *Encyclopedia Britannica*. Great Books # 42, page 295. January 1834.

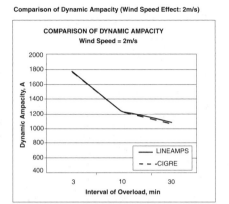

FIGURE 4.9 Ampacity calculated by LINEAMPS program in dynamic state is compared to Cigré method for the calculation of dynamic ampacity when wind speed is 2 m/s.

TABLE 4.4
Data for Line Ampacity Calculations Presented in Figures 4.4–4.9

Transmission Line Conductor	795 kcmil 26/7 ACSR Drake
Wind Speed	2 ft/s
Wind Direction	Perpendicular to Line
Latitude	30 °
Azimuth of Conductor	90°
Atmosphere	Clear
Solar Heating	On
Diffuse Solar Radiation	0 (ignored in IEEE & Cigré), considered in LINEAMPS
Emissivity	0.5
Absorptivity	0.5
Elevation above Sea Level	0 m
Ground Surface Type	Urban
Time of Day	11:00 am
Time of Year	June 10
Maximum Conductor Temperature	100°C

Source: N.P. Schmidt, Comparison between IEEE and Cigré ampacity standards, IEEE Power Engineering Society conference paper # PE-749-PWRD-0-06-1997. Anjan K. Deb, Discussion contribution, October 1997.

The same quantity of electricity which, passed in a given time, can heat an inch of platina wire of a certain diameter red-hot can also heat a hundred, a thousand, or any length of the same wire to the same degree, provided the cooling circumstances are the same for every part in all cases.

Experimental Verification of Transmission Line Ampacity 71

In an overhead power transmission line the cooling effects are generally not the same at all sections of the line because of its length. A transmission line may be 10 or 100 miles long (or greater), and the meteorological conditions cannot be expected to remain the same everywhere. LINEAMPS takes into consideration the different cooling effects on the transmission line conductor, due to varying meteorological conditions in space and in time, by object-oriented modeling of transmission lines and weather stations, by introducing the concept of virtual weather sites,* and by expert rules described in Chapter 8.

4.4.2 Dynamic Ampacity

In the dynamic state, short-term overload currents greater than steady-state ampacity are allowed on a transmission line by taking into consideration the energy stored in a transmission line conductor. The energy stored in a transmission line conductor is shown by the differential equation (3.5) in Chapter 3. In addition, Section 3 of a recent Cigré report** presents data on dynamic ampacity. The data was compared to the values calculated by a LINEAMPS Dynamic model with excellent agreement, as shown in Figures 4.8 and 4.9.

4.5 MEASUREMENT OF TRANSMISSION LINE CONDUCTOR TEMPERATURE

345 kv Transmission line

The results obtained from the LINEAMPS program were also verified by comparison with the ampacity of a real transmission line by measurement. Measurements were made by temperature sensors installed on various locations of a 345 kV overhead transmission line operated by the Commonwealth Edison Company (ComEd) in the region of Chicago, IL. The results of this comparison are presented in Figure 4.10, showing that the ampacity of the transmission line calculated by the LINEAMPS program never exceeded the measured values at all locations during daytime for the period considered in the study. These results clearly indicate that the program safely and reliably offers substantial increase in line ampacity over the present method of static line rating.

As seen in Figure 4.10, LINEAMPS ratings never exceeded measured (ComEd) ampacity at different hours of the day. It also accurately predicted the lowest value of line ampacity at noontime. The ampacity predicted by LINEAMPS offers substantially higher line capacity than the present method of static line rating. The static line ampacity is 1000 A, as shown in Figure 4.10.

---

* LINEAMPS User Manual, 1998.
** The thermal behaviour of overhead conductors. Section 3: Mathematical model for evaluation of conductor temperature in the unsteady state. Cigré Working Group 22.12 Report, *Electra*, October 1997.

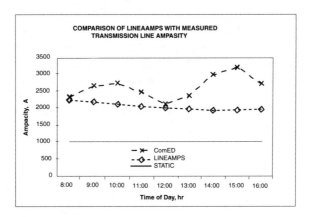

FIGURE 4.10 Transmission line ampacity calculated by LINEAMPS program is compared to the ampacity measured on a real 345 kV overhead transmission line operated by Commonwealth Edison Company in the region of Chicago, IL, USA.

4.6 CHAPTER SUMMARY

In this chapter the results calculated by the LINEAMPS program are compared to experimental data from different power company data, as well as transmission line conductor manufacturers' catalog data. The results of these comparisons show that line ampacity calculated by program is in good agreement with actual data from the field. The calculations are also compared to data presented at the IEEE and Cigré conferences by several researchers. In all of these comparisons, there is excellent agreement with the results obtained by program. Therefore, according to Faraday, the proposed theory of transmission line ampacity, conductor thermal models, hypotheses, and the correctness of various assumptions are validated by verifying results obtained by program with experimental data.

5 Elevated Temperature Effects

5.1 INTRODUCTION

The advantages of higher transmission line ampacity discussed in Chapter 1 include the deferment of the capital investment required for the construction of new lines and economic energy transfer. As a result of achieving higher line ampacity, electricity costs are reduced and there is less environmental impact. While there are significant benefits to increasing transmission line ampacity, its effects must be clearly understood and evaluated accurately. In this chapter the effects of higher transmission line ampacity are evaluated from the point of view of elevated temperature operation of conductors. The problem of electric and magnetic fields due to higher ampacity are presented in Chapter 6. This chapter includes a study of transmission line conductor sag and tension, permanent elongation, and the loss of tensile strength of the powerline conductor due to elevated temperature operation.

As stated in the previous chapters, the main objective of the powerline ampacity system is to accurately predict transmission line ampacity based upon actual and forecast weather conditions. The line ampacity system will ensure that the allowable normal and emergency operating temperatures of the conductor are not exceeded. The line ampacity system program should also verify that the loss of tensile strength of a conductor is within acceptable limits, and that any additional conductor sag caused by permanent elongation will not exceed design sag and tension during the lifetime of the transmission line conductor. When line ampacity is increased conductor temperature increases, consequently, there may be greater loss of tensile strength of conductor and higher sag, which must be evaluated properly.

The loss of tensile strength and permanent elongation of a conductor is calculated recursively from a specified conductor temperature distribution by using the empirical equations found in the literature (Harvey, 1972; Morgan, 1978; Cigré, 1978; Deb et al., 1985; Mizuno, 1998). The Cigré report did not describe how the temperature distribution was obtained, and did not include the design sag and tension of conductor. An elegant method to calculate sag and tension by a strain summation procedure is described in a report prepared by Ontario Hydro.* The unique contribution made in this chapter is the development of an unified approach to determine sag and tension during the lifetime of a transmission line conductor by consideration of the probability distribution of transmission line conductor temperature.

---

* Development of an accurate model of ACSR conductors at high temperatures. Canadian Electricity Association Research Report.

5.1.1 EXISTING PROGRAMS

Most sag-tension computer programs presently used are based on the assumption that conductor temperature will remain constant for the entire life of the line. In reality, as we all know, conductor temperature is never constant. For example, conductor sag at 100 °C cannot be expected to remain the same if it has been operated at that temperature for 100 hours or 10,000 hours. Therefore, sag based on the probability distribution of conductor temperature is required. Conductor sag and tension are important transmission line design parameters upon which depend the security of the line. A line security analysis was carried out (Hall, Deb, 1988) based upon different line operating conditions. This study showed how conductor sag and tension varies with conductor temperature frequency distributions.

5.2 TRANSMISSION LINE SAG AND TENSION — A PROBABILISTIC APPROACH

A method of calculation of conductor sag and tension is presented in this section by consideration of the probability distribution of transmission line conductor temperature in service. The probability distribution of conductor temperatures is obtained by the synthetic generation of meteorological data from time-series stochastic and deterministic models. This method of generating probability distribution of conductor temperatures takes into account the correlation between the meteorological variables and the transmission line current.*

The effects of elevated temperature operation of conductors comprising inelastic elongation and the loss of tensile strength of conductor are considered by the recursive formulation of inelastic elongation and annealing models found in the literature. The equations and algorithm that were used to calculate conductor sag and tension from the probability distribution of conductor temperatures are presented and implemented in a computer program. Results are presented that show good agreement with data from other computer programs.

There is considerable interest in the industry in the probabilistic design of overhead lines.** Ghanoum (1983) described a method for the structural design of transmission lines based upon probabilistic concepts of limit loads and return period of wind. Probability-based transmission line rating methods are described by several authors (Koval and Billinton, 1970; Deb et al., 1985, 1993; Morgan, 1991; Redding, 1993; Urbain, 1998). Redding*** 1993 presented probability models of ambient temperature and wind speed, but the resulting conductor temperature distribution was not given. Not much attention has been given to probabilistic design of sag and tension of overhead line conductors, which depend upon conductor temperature probability distributions. The probability distribution of conductor temperature is a

---

* See discussion contribution by J.F. Hall and Anjan K. Deb on the IEEE paper (Douglass, 1986).
** Ghanoum, E., "Probabilistic Design of Transmission Lines," Part I, II. *IEEE Transactions on Power Apparatus and Systems*, Vol. PAS-102, No. 9, 1983.
*** J.L. Redding, "A Method for Determining Probability Based Allowable Current Ratings for BPA's Transmission Lines, "IEEE/PES 1993 Winter Meeting Conference Paper # 93WM 077-8PWRD, Columbus, Ohio, January 31–February 5, 1993.

Elevated Temperature Effects

function of conductor current and the meteorological conditions on the line (Hall and Deb, 1988; Morgan, 1991; Mizuno et al., 1998, 2000). Mizuno et al. considered the loss of strength of conductor as the index of thermal deterioration. In this chapter I have considered transmission line sag and tension as the determining factor. This includes both the loss of strength and the permanent elongation of the conductor.

5.2.1 THE TRANSMISSION LINE SAG-TENSION PROBLEM

Given the maximum mechanical loading of a conductor due to wind and ice, and a probable distribution of conductor temperature with time, it is required to calculate the sag and tension of the conductor at different temperatures. The probability distribution of conductor temperature is obtained from line ampacity simulations. Transmission line sag and tension are critical line design parameters that are required to verify conductor-to-ground clearance and the safety factor of the conductor at the maximum working tension.

5.2.2 METHODOLOGY

In order to predict conductor sag at the highest conductor temperature, the permanent elongation of the conductor due to metallurgical creep is required to be estimated. This requires an estimate of the conductor temperature distribution during the expected life of the line. Future conductor temperature distributions require knowledge of the line current as well as the meteorological conditions. The conductor temperature distribution is also required to calculate the loss of strength of the conductor.

Future meteorological conditions may be estimated by the random generation of weather data from their specified probability distributions, or by taking a typical set of weather data and assuming it to repeat itself every year (Giacomo et al., 1979). In the method proposed by the author (Deb, 1993), meteorological data is generated by Monte Carlo simulation of the following time series stochastic and deterministic model:

$$Y(t) = X(t)^T \cdot A(t) + \eta(t) \tag{5.1}$$

$Y(t) \in (Ta, Nu, Sr)$ input variables*
$X(t)^T = \{1, Sin(\omega t), Sin(2\omega t), Cos(\omega t), Cos(2\omega t), z(t-1), z(t-2)\}$
$A(t)$ = model coefficients
ω = fundamental frequency
$z(t-1), z(t-2)$ are the stochastic variables at lag 1 and lag 2 respectively
$\eta(t)$ = uncorrelated white noise

By this method it is possible to take into consideration time-of-day effects of weather and line current in the analysis. Using real weather data in chronological order allows

---

* Ta = ambient temperature, Nu = Nusselt number (coefficient of heat transfer), Sr = solar radiation (Sr is also calculated analytically by the method given in Chapter 7).

this also. Time-of-day effects are lost when weather data is generated from known probability distributions (Mizuno et al., 1998), (Deb et al., 1985).

AAC Bluebell conductor temperature distribution is shown in Figure 5.5 which was obtained by the synthetic generation of California meteorological data from time series stochastic models. Examples of synthetic generations of meteorological data from time series stochastic models are shown in Figures 5.1–5.4.

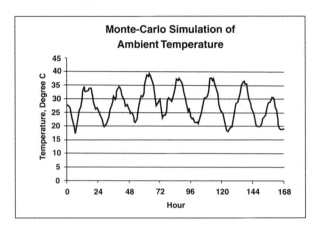

FIGURE 5.1 Hourly values of ambient temperature generated by Monte Carlo simulation for the San Francisco Bay area.

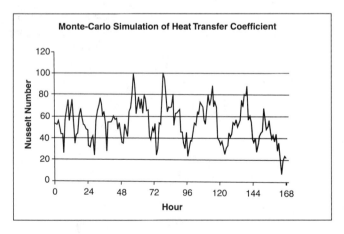

FIGURE 5.2 Hourly values of conductor heat transfer coefficient of AAC Bluebell transmission line conductor generated by Monte Carlo simulation for the San Francisco Bay area.

The conductor temperature distribution of Figure 5.5 assumes constant line current equal to the static line rating and is used to calculate transmission line conductor sag and tension.

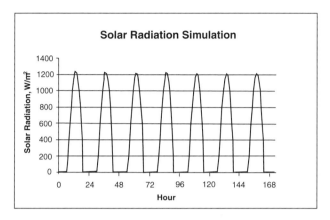

FIGURE 5.3 Hourly values of solar radiation on conductor surface of AAC Bluebell transmission line conductor in the San Francisco Bay area simulated by program.

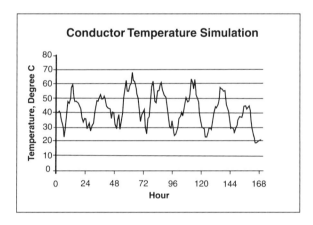

FIGURE 5.4 Hourly values of conductor temperature of AAC Bluebell transmission line conductor in the San Francisco Bay area using simulated weather data.

5.2.3 Computer Programs

The following computer programs are required for the prediction of sag and tension at high temperature based upon the probabilistic distribution of conductor temperature:

1. Conductor temperature predictor
2. Probability distribution generator of ambient temperature, line current, solar radiation, and conductor heat transfer coefficient from time-series stochastic models
3. Probability distribution generator of conductor temperatures
4. Inelastic elongation (creep) predictor
5. Loss of strength predictor
6. Sag and Tension Calculator

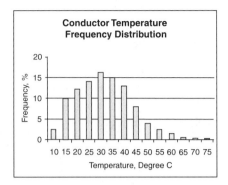

FIGURE 5.5 Frequency distribution of AAC Bluebell transmission line conductor temperature in the San Francisco Bay area simulated by program.

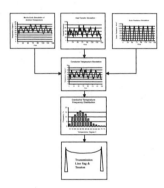

FIGURE 5.7 Flow chart for the calculation of transmission line conductor sag and tension from probability distribution of conductor temperature.

The above modules constitute the Sag and Tension Program developed by the author. A flow chart for the calculation of transmission line sag and tension from the probability distribution of conductor temperature is given in Figure 5.7. The equations and algorithms are developed in the following sections.

5.3 CHANGE OF STATE EQUATION*

The change of state equation given below is used for the calculation of transmission line sag and tension. If the conductor is at State 1 given by conductor stress σ_1 and temperature Tc_1 and goes to State 2 given by stress σ_2 and temperature Tc_2, then the new sag and tension of the conductor at State 2 is calculated from the following change of state equation:

---

* P. Hautefeuille, Y. Porcheron, Lignes Aeriennes, *Techniques de l'Ingenieur*, Paris.
J.P. Bonicel, O. Tatat, Aerial optical cables along electrical power lines, *REE* No. 3, March 1998, SEE France.

Elevated Temperature Effects

$$\frac{\sigma_2}{E} - \frac{(\varpi \cdot L)^2}{24\sigma_2^2} + \alpha(Tc_2 - Tc_1) + \Delta Ec = \frac{\sigma_1}{E} - \frac{(\varpi \cdot L)^2}{24\sigma_1^2} \qquad (5.2)$$

σ_1, σ_2 = stress at State1 and State2 respectively, kg/mm²
Tc_1, Tc_2 = conductor temperature at State1 and State2, °C
E = Young's modulus of elasticity, kg/mm²
ϖ = specific weight of conductor, kg/m/mm²
L = span length, m
ΔEc = inelastic elongation (creep) mm/mm
α = coefficient of linear expansion of conductor, °C⁻¹

Conductor sag is calculated approximately by the following well-known parabola equation:

$$Sag = \frac{WL^2}{8T} \qquad (5.3)$$

where,

Sag is in meters, m
W = conductor weight, kg/m
T = conductor tension, kg

The above equation is used to calculate transmission line conductor sag and tension. It requires a knowledge of Young's modulus of elasticity, the coefficient of linear expansion of the conductor and the permanent elongation of the conductor due to elevated temperature creep. Young's modulus is obtained from the stress/strain curve shown in Figure 5.6. The coefficient of linear expansion is a property of the conductor. The elevated temperature creep Ec (metallurgical creep) is estimated separately by the creep predictor program.

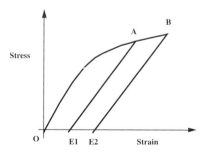

FIGURE 5.6 Stress/strain relation of AAC conductor

5.3.1 RESULTS

The probability distribution of conductor temperature is shown in Table A5.1 in Appendix 5 at the end of this chapter, the loss of strength is given in Table A5.2, and the permanent elongation of conductor during the lifetime of a transmission line conductor is presented in Table A5.3. Calculations are based upon the conductor temperature distributions generated from Figure 5.5. The sag and tension of transmission line conductors is presented in Table A5.6 of Appendix 5. Comparison of sag and tension results with other programs are given in the Tables A5.8 through A5.10. Further results of transmission line sag and tension of special conductors are given in Table A5.11 (Choi et al., 1997). The study was prepared by the author with KEPCO* for line uprating with high-ampacity conductors. The results obtained by program show excellent agreement with field measurements on an actual transmission line in Korea.

5.3.2 Conductor Stress/Strain Relationship

The stres/strain relationship of an all aluminum conductor is shown in the Figure 5.6 for the purpose of illustration of some of the concepts discussed in this section. The stress/strain relationship of an ACSR conductor is somewhat complicated, though the general concepts remain the same for any type of conductor.

When tension is applied to an unstretched conductor, the ratio stress/strain of the conductor follows the curve OA. When the tension is lowered at Point A, this ratio becomes linear and follows the trace AE1. If the tension is increased again it follows the linear path E1A until it reaches Point A. If the tension is further increased at Point A it becomes nonlinear again, as shown by the curve AB. When tension is lowered again at Point B, it follows the linear path BE2. In Figure 5.6, the section of the stress/strain curve OA and AB represents the initial stress/strain curve, which is nonlinear. Consequently, the Young's modulus in this region becomes nonlinear and is generally approximated by fitting a polynomial function of degree N to the data. The final modulus of elasticity given by the slope of the linear portion of the curves AE1 and BE2 is constant.

The sections OE1 and E1E2 are the permanent elongation due to creep (geometrical settlement). The advantage of pretensioning the conductor becomes obvious from Figure 5.6. The permanent stretch OE1 and E1E2 can be removed if the conductor is pretensioned by a load to reach Point B close to the allowed maximum working tension of the line. The stress/strain curve then becomes linear.

5.4 PERMANENT ELONGATION OF CONDUCTOR

Permanent or irreversible elongation of the conductor is known to occur due to elevated temperature operation of the conductor. It results in increase of conductor sag and reduces the midspan clearance to ground. Elevated temperature creep is a function of the conductor temperature, its duration, and the conductor tension. Two factors cause permanent elongation of the conductor (Cigré, 1978):

1. Geometric settlement
2. Metallurgical creep

---

* Korea Electric Power Company.

5.4.1 GEOMETRIC SETTLEMENT

Geometric settlement depends upon conductor stringing tension and occurs very rapidly as it only involves the settling down of strands. Generally, the process starts with conductor stringing and is completed within 24 hours. Higher than normal stringing tensions, "pretension," is sometimes applied to a conductor to accelerate the process of geometric settlement. The geometric settlement, Es, is calculated by the following formula (Cigré, 1978):

$$Es = 750(d - 1)(1 - \exp(-m/10))(MWT/RTS)^{2.33} \tag{5.4}$$

d = wire diameter, mm
m = aluminum/steel sectional area, ratio
MWT = Maximum Working Tension, kg
RTS = Rated Tensile Strength, kg

5.4.2 METALLURGICAL CREEP

Metallurgical creep is a function of conductor temperature, tension, and time. Therefore, elevated temperature operation of a line for short duration is not as much of a concern as continuous operation at high temperatures. Metallurgical creep is estimated by using the following empirical formula determined experimentally (Cigré, 1978):

$$E_c = \frac{1}{\cos^{2+\alpha}\beta} K \cdot \exp(\phi T_c)\sigma^\alpha t^{\mu/\sigma^\delta} \tag{5.5}$$

E_c = elongation, mm/km
$K, \phi, \alpha, \mu, \delta$ are constants (Table 5.1 gives values for typical ACSR conductor sizes)
Tc = average conductor temperature, °C
σ = average conductor stress, kg/mm²
t = time, hr

The factor β takes into account the effect of conductor type, stranding and material and is calculated as follows,

$$\beta = \frac{\sum_{i=1}^{N} n_i \beta_i}{\sum_{i=1}^{N} n_i} \tag{5.6}$$

N = number of aluminum strands
n_i = number of wires in layer i
β_i = angle of the tangent in a point of a wire in layer i with conductor axis

5.4.3 RECURSIVE ESTIMATION OF PERMANENT ELONGATION

The inelastic elongation of a conductor due to metallurgical creep, E_c, is calculated in small steps, ΔE_c, by the following recursive equations:

$$Ec_{i,j} = Ec_{i,j-1} + \Delta Ec \tag{5.7}$$

$$tq_{i,j} = \left\{ Ec_{i,j-1} K' \sigma_{i,j}^{-\alpha} \exp(-\phi Tc_i) \right\}^{-\mu/\sigma^\delta} \tag{5.8}$$

$i = 1,2\ldots n$ line loading intervals obtained from the frequency distribution of Table A5.1

TABLE 5.1
Value of coefficients in equation (5.5) (Cigré, 1978)

ACSR Al/St	K	f	a	m	d
54/7	1.1	0.0175	2.155	0.342	0.2127
30/7	2.2	0.0107	1.375	0.183	0.0365

$j = 1,2\ldots f$ sub intervals
$tq_{i,j}$ = equivalent time at present temperature Tc_i for the past creep $Ec_{i,j-1}$

$$K' = \frac{\cos^{2+\alpha}\beta}{K} \tag{5.9}$$

Total creep is estimated by summation,

$$E_{total} = Ec_{n,f} + Es_f \tag{5.10}$$

$Ec_{n,f}$ and Es_f are the final inelastic elongation due to metallurgical creep and geometrical settlement of the transmission line conductor.

Results are presented in Table A5.3 of Appendix 5 for the frequency distribution of Table A5.1. When Table A5.3 values are used to calculate sag, the results are given in Table A5.6. When final sags are compared to the initial conditions specified in Table A5.5, it is seen that the sag of the AAC Bluebell conductor increases by 5.9 ft (14.4% increase) over initial conditions after 30 years. When maximum conductor temperature is 75°C, the increase in sag is only 10.25% after 30 years. Transmission line design conditions will determine whether the increase in sag or loss of strength is the limiting factor for a particular transmission line.

Elevated Temperature Effects

5.5 LOSS OF STRENGTH

The loss of tensile strength of a conductor results in lowering the design safety factor of the conductor. Generally, T/L conductor tension is designed with a safety factor of two at the worst loading condition.* The worst condition results in the conductor being subjected to the maximum tension. Such conditions arise when the conductor is exposed to high winds and/or ice covering. To give an example, an ACSR Cardinal conductor under extreme wind (~ 100 mph) and ice loading may result in the tension of the Cardinal conductor to reach 17,000 lbf which is approximately 50% of the rated tensile strength of the conductor. Therefore, a 10% reduction in the tensile strength of the conductor would also lower the safety factor of the conductor by 10%. Generally, a loss of strength up to 10% is acceptable (Mizuno et al., 1998).

5.5.1 PERCENTILE METHOD

A recent study (Mizuno et al., 1998) describes the calculation of thermal deterioration of a transmission line conductor by a probability method. The reduction in tensile strength of the conductor was used as the index of thermal deterioration. The loss of tensile strength is calculated as a function of conductor temperature, Tc, and the time duration, t, at which the temperature, Tc, is sustained (Morgan, 1978); (Harvey, 1972).

$$W = \exp\{C(\ln t - A - BT)\} \quad (5.11)$$

W is loss of strength (%) and A,B,C are constants that are characteristics of the conductor. This is an empirical equation based upon laboratory tests on individual wire strands. The total loss of strength is then obtained by,

$$\sum W = \left[\left(\ldots \left((t_1 \bar{t}_2/\bar{t}_1) + t_2 \right) \bar{t}_3/\bar{t}_2 + \cdots t_{n-1} \right) \bar{t}_n/\bar{t}_{n-1} \right) + t_n \right) \bar{t}_n \right]^c \quad (5.12)$$

where t_i is time duration when the conductor temperature is T_i and $\bar{t}_i$ is given by,

$$\ln \bar{t}_i = A + BT \ln \quad (5.13)$$

5.5.2 RECURSIVE ESTIMATION OF LOSS OF STRENGTH

The author has developed a recursive method for the calculation of loss of strength of conductors as follows:

$$W = W_a[1 - \exp\{-\exp(A_3 + B_3 T_c + n_1 \ln t + K \ln (R/80))\}] \quad (5.14)$$

$$Wi = Wa[1 - \exp\{-\exp(A3 + B3Tci + n_1 \ln(ti + tqWi-1) + K\ln (R/80))\}] \quad (5.15)$$

---

* National Electrical Safety Code C2-1997.

$$tqW_{i-1} = \exp[\ln \ln\{1/(1 - W_{i-1}/W_a)\} + A_3$$
$$+ B_3 T_{ci} + n_1 \ln(t_i + tqW_{i-1}) + K\ln(R/80)\}]/n_i \quad (5.16)$$

A_3, B_3, n_1, K = constants given in Table 5.2 (Morgan. 1978).

TABLE 5.2
Value of Coefficients in Equations (5.15), (5.16) (Morgan 1978)

Wire	A_3	B_3	n_1	K
Aluminum	−8.3	0.035	0.285	9
Aluminum Alloy	−14.5	0.060	0.79	18
Copper	−7.4	0.0255	0.40	11

R = Reduction of wire by drawing from rod to wire (Morgan, 1978).

$$R = 100\left[1 - \frac{D}{D_o}\right]^2 \quad (5.17)$$

D, D_o are the diameters of wire and rod, respectively
i = 1, 2, 3....n intervals of time
tqW_{i-1} = equivalent time for loss W_{i-1} at temperature Tc_i

Results are presented in Appendix 5 in Table A5.2 for the frequency distribution of Table A5.1. These results show that the loss of tensile strength for the AAC Bluebell conductor is greater than 10% when the maximum temperature of the conductor is 95°C and the life is 25 years. For this reason, All Aluminum Conductors (AAC) are generally operated below 90°C under normal conditions.*

5.6 CHAPTER SUMMARY

When transmission line ampacity is increased it is necessary to properly evaluate the thermal effects of the powerline conductors, which includes loss of tensile strength of the conductor, permanent elongation, and conductor sag. In this chapter a unified approach to modeling and evaluation of the elevated temperature effects of transmission line conductors is presented. Conductor loss of strength and permanent elongation are evaluated recursively from the probability distribution of conductor temperature. The probability distribution of conductor temperature is generated by Monte Carlo simulation of weather data from time-series stochastic models and transmission line current. Therefore, a new method is developed to determine the sag and tension of overhead line conductors with elevated temperature effects.

---

* PG&E Line Rating Standard.

A study of the AAC Bluebell conductor is presented to show the long-term effects of elevated temperature. Results are presented which clearly show that the loss of strength of the conductor is less than 10% when the maximum temperature of the conductor is 90°. The increase in sag is less than 15% for the same maximum operating temperature.

The accuracy of the sag-tension program was tested with ALCOA's Sag and Tension program, Ontario Hydro's STESS program, and KEPCO's transmission line field data. Results are presented that compare well with all of the above data.

Appendix 5
Sag and Tension Calculations

FREQUENCY DISTRIBUTION OF CONDUCTOR TEMPERATURE

A conductor temperature distribution is shown in Table A5.1. This temperature distribution was generated by assuming constant load current equal to the static line rating, and by the artificial generation of meteorological data. From the basic temperature distribution assumption shown in Column 1 of Table A5.1, the conductor temperatures were increased in steps of 5°C up to the maximum temperature of 100°C. The corresponding frequency distributions of conductor temperature are also shown in this table.

TABLE A5.1
Frequency Distribution of Conductor Temperature

Maximum Conductor Temperature, °C						
75	80	85	90	95	100	Frequency, %
10	15	20	25	30	35	2.0
15	20	25	30	35	40	11.0
20	25	30	35	40	45	13.0
25	30	35	40	45	50	14.0
30	35	40	45	50	55	16.0
35	40	45	50	55	60	15.0
40	45	50	55	60	65	12.5
45	50	55	60	65	70	8.0
50	55	60	65	70	75	4.0
55	60	65	70	75	80	2.0
60	65	70	75	80	85	1.5
65	70	75	80	85	90	0.5
70	75	80	85	90	95	0.3
75	80	85	90	95	100	0.2
						Σ 100%

Table A5.1 provides the starting point in the calculation of loss of strength and permanent elongation of conductor. The sag and tension of the conductor are then calculated from this data. The factor of safety of the conductor and the line to ground clearances can now be verified.

Loss of Tensile Strength of Conductor

Based upon the distributions shown in Table A5.1, the loss of strength of the conductor was calculated at different conductor maximum temperatures for the conductor life, from 10 to 30 years. The results are shown in Table A5.2 for the AAC Bluebell conductor.

TABLE A5.2
Loss of Strength of AAC Bluebell Calculated by Program

Maximum Conductor Temperature °C	Conductor Life, Year				
	10	5	20	25	30
	Loss of Strength, %				
75	4.3	4.8	5.2	5.5	5.8
80	5.1	5.6	6.1	6.5	6.8
85	6.0	6.7	7.2	7.6	8.0
90	7.0	7.8	8.5	9.0	9.4
95	8.3	9.2	9.9	10.5	11.0
100	9.7	10.8	11.6	12.3	12.9

Permanent Elongation of Conductor

The permanent elongation of the AAC Bluebell conductor due to metallurgical creep calculated by the program is shown in Table A5.3 and is based on the same assumptions of conductor temperature distribution as shown in Table A5.1.

TABLE A5.3
Permanent Elongation of AAC Bluebell Calculated by Program

Maximum Conductor Temperature °C	Life of Conductor, Year				
	10	15	20	2	30
	Permanent Elongation, Micro Strain				
75	809	859	899	939	969
80	889	959	999	1049	1079
85	989	1069	1119	1169	1209
90	1109	1189	1249	1289	1329
95	1239	1309	1369	1419	1459
100	1359	1449	1529	1579	1619

The loss of strength and permanent elongation of the conductor shown in the Tables A5.2 and A5.3, respectively are used as input data to the sag and tension program. The sag and tension of the 1034 Kcmil AAC Bluebell is given in Table A5.4.

Appendix 5 Sag and Tension Calculations

SAG AND TENSION CALCULATION BY PROGRAM

TABLE A5.4
Input Data

1034 Kcmil AAC Bluebell

Conductor diameter, in	1.17
Mass, lb/ft	0.97
Rated Tensile Strength, lbf	18500
Modulus of elasticity, psi	$8.5 \cdot 10^6$
Coefficient of linear expansion, /°C	$23 \cdot 10^{-6}$
Sectional area, in^2	0.81
Final unloaded tension, lbf	3700
Final unloaded temperature, °F	50

TABLE A5.5
Sag and Tension Initial Condition after Sagging-In

Span ft	Wind lb/in2	Ice in	Tc °F	Tc °C	Tension lb	SF #	Sag ft
1000	8.0	0	25	−3.9	4972	3.7	31.9
1000	0	0	60	15.6	3641	4.9	33.4
1000	0	0	167	75.0	3045	5.8	40.0
1000	0	0	176	80.0	3006	5.9	40.5
1000	0	0	185	85.0	2969	5.9	41.0
1000	0	0	194	90.0	2934	5.9	41.5
1000	0	0	203	95.0	2899	5.9	42.0
1000	0	0	212	100.0	2866	5.9	42.5

TABLE A5.6
Final Sag

Temperature °C	Life, year				
	10	15	20	25	30
			Sag, ft		
75	43.4	43.6	43.8	44.0	44.1
80	44.2	44.5	44.7	44.9	45.0
85	45.1	45.4	45.6	45.8	46.0
90	46.0	46.4	46.6	46.7	46.9
95	47.0	47.3	47.5	47.7	47.8
100	47.9	48.2	48.5	48.7	48.9

Sag and Tension Comparison with STESS Program of Ontario Hydro

To verify the accuracy of the results obtained from the sag and tension program, it was compared to the STESS program. The results are provided below.

TABLE A5.7
Input

ACSR Drake

Conductor diameter, mm	28.13
Mass, kg/m	1.628
Rated Tensile Strength, kN	140.1
Modulus of elasticity, kg/mm	8360
Coefficient of linear expansion /°C	$19 \cdot 10^{-6}$
Sectional area, mm^2	468.7
Final unloaded tension, kN	27.80
Final unloaded temperature, °C	20

TABLE A5.8
Sag and Tension Comparison with STESS Program of Ontario Hydro

Span ft	Wind Pa	Tc °C	Tension kN	Safety Factor, #	Sag, m STESS	Sag, m Program*
300	0	20	28.0	5.0	6.47	6.4
300	0	30	26.4	5.3	6.84	6.8
300	0	40	25.1	5.6	7.20	7.2
300	0	50	23.8	5.9	7.56	7.5
300	0	60	22.8	6.1	7.92	7.9
300	0	70	21.8	6.4	8.27	8.2
300	0	80	21.0	6.7	8.62	8.6
300	0	90	20.2	6.9	8.96	8.9
300	0	100	19.5	7.2	9.27	9.3
300	0	110	18.8	7.4	9.45	9.5
300	0	120	18.3	7.7	9.64	9.6
300	0	130	17.7	7.9	9.82	9.8
300	0	140	17.2	8.1	10.01	9.9
300	0	150	16.8	8.4	10.20	10.1

Note:
1 Pa = 0.02 lb/ft^2
1 kN = 225.8 lbf
Wind = Wind Pressure on Projected Area of Conductor, Pa
Tc = Average Conductor Temperature, °C
*Calculated by the sag and tension program described in this chapter.

Appendix 5 Sag and Tension Calculations

SAG AND TENSION COMPARISON WITH ALCOA PROGRAM*

Results obtained by new program are compared to the ALCOA Sag and Tension Program (see footnote).

TABLE A5.9
Input (ALCOA Program Data)

795 AS33 ACSR DRAKE (26 al + 7 st)

Conductor diameter, in	1.108
Mass, lb/ft	1.0940
Rated Tensile Strength, lb	31500
Modulus of elasticity, psi	$11.89 \cdot 10^6$
Coefficient of linear expansion, /°C	$19.5 \cdot 10^{-6}$
Sectional area, in^2	0.7264
Final unloaded tension, lbf	4403
Final unloaded temperature, °F	60

TABLE A5.10
Sag and Tension Comparison with ALCOA Program

Span ft	Wind lb/ft2	T °Fc	Tension lbf	Safety Factor, SF	Sag, ft ALCOA	Sag, ft Program*
750	33.0	60	11496	2.7	21.8	21.9
750	0	70	4293	7.1	18.0	18.0
750	0	80	4171	7.3	18.5	18.5
750	0	90	4057	7.5	19.0	19.0
750	0	100	3951	7.7	19.5	19.5
750	0	110	3852	7.9	20.0	20.0
750	0	120	3784	8.1	20.5	20.3
750	0	170	3578	9.0	21.5	21.5
750	0	205	3450	9.7	22.3	22.3

Wind = Wind Pressure on Projected Area of Conductor, lb/ft^2
Tc = Average Conductor Temperature, °F
* Calculated by the sag and tension program described in this chapter.

---

* Craig B. Lankford, ALCOA's Sag and Tension Program Enhanced for PC Use, *Transmission and Distribution Journal*, Vol. 41, No. 11, November 1989.

TABLE A5.11
Sag and Tension Comparison with KEPCO Line Data

154 kV Double Circuit Line

INPUT	ACSR	STACIR
Conductor dia, mm	25.30	25.30
Mass, kg/m	1.30	1.30
Rated Tensile Strength, kN	98.00	98.00
Modulus of Elasticity, kg/mm^{2}mm	8360.00	16500.00
Coeff. of lin. expan. /deg C	19E-06	3GE-07
Sectional Area, mm^{2}mm	379.60	379.60
Final unloaded tension, kN	24.50	24.50
Final unloaded temperature, /deg C	10.00	10.00

OUTPUT			ACSR		STACIR	
Span, m	Wind Pa	Tc Deg C	Tension kN	Sag m	Tension kN	Sag m
300	400	10	29.2	6.3	29.2	6.3
300	0	10	24.5	5.9	24.5	5.9
300	0	90	16.4	8.8	18.0	8.2
300	0	150	NA	NA	17.1	8.4
300	0	200	NA	NA	16.4	8.7
300	0	210	NA	NA	16.3	8.8
300	0	240	NA	NA	16.0	9.0

NA = Not Applicable

Table adapted from KEPCO high-ampacity transmission line (Choi et al., 1997).

6 Transmission Line Electric and Magnetic Fields

6.1 INTRODUCTION

The magnetic field of a transmission line increases with line ampacity, and increases at ground level with conductor sag. Typical powerline configurations are evaluated to show their magnetic fields. It is shown that the magnetic field of overhead transmission lines is within acceptable limits. Line designs and EMF mitigation methods are developed to lower transmission line magnetic fields in sensitive areas. Electric field limit at ground level is not exceeded by higher transmission line ampacity if the maximum design temperature of the conductor is not exceeded and minimum conductor-to-ground distance is maintained.

Even though transmission line voltage remains unchanged with higher ampacity, electric fields are also evaluated since the level of an electric field at ground is affected by conductor sag. It must be mentioned that the lowering of transmission line conductor to ground distance due to higher sag will raise the level of electric and magnetic fields at ground level. Therefore, it is important to calculate transmission line sag accurately for the calculation of electric and magnetic fields at ground level.

The study of transmission line magnetic fields is also important from the point of view of transmission line ampacity. In the previous chapters it was shown that transmission line capacity may be increased by dynamic line ratings, which will result in the lowering of the cost of electricity. On the other hand, increasing line ampacity also increases the level of magnetic field. With public concern for electric and magnetic fields, transmission line engineers are required to accurately evaluate the impact of increased line capacity on the environment due to higher electric and magnetic fields.

6.2 TRANSMISSION LINE MAGNETIC FIELD

The magnetic field of a current-carrying transmission line conductor is calculated by the application of Maxwell's equation. The Electric Power Research Institute (EPRI) conducted a study* in which they proposed methods for the reduction of transmission line magnetic fields. The study presents data to quantify more accurately the magnetic fields of different transmission line configurations.

---

* V.S. Rashkes, R. Lordan, "Magnetic Field Reduction Methods: Efficiency and Costs," *IEEE Transactions on Power Delivery*, Vol. 13, No. 2, April 1998.

In this section, a general method of calculation of the magnetic field of overhead transmission line conductors is presented. This method is suitable for any transmission line configuration. The magnetic fields of typical transmission line configurations are also presented to show that magnetic fields are within acceptable limits. It is shown that there is minimum environmental impact due to higher transmission line ampacity. EMF mitigation methods are also given.

6.2.1 THE MAGNETIC FIELD OF A CONDUCTOR

In the following section we shall derive expressions for the calculation of the magnetic field of a current-carrying conductor by the application of Maxwell's equations.

$$(\nabla \times H)_r = \frac{1}{r} \cdot \frac{\partial Hz}{\partial \varphi} - \frac{\partial H\varphi}{\partial z} = J_r \qquad (6.1)$$

$$(\nabla \times H)_\varphi = \frac{1}{r} \cdot \frac{\partial Hz}{\partial z} - \frac{\partial Hz}{\partial r} = J_\varphi \qquad (6.2)$$

$$(\nabla \times H)_z = \frac{1}{r} H\varphi + \frac{\partial H\varphi}{\partial r} - \frac{\partial Hr}{\partial \varphi} = J_z \qquad (6.3)$$

$$\nabla \cdot H = \frac{1}{r} Hr + \frac{\partial Hr}{\partial r} + \frac{1}{r} \cdot \frac{\partial H\phi}{\partial \phi} + \frac{\partial Hz}{\partial z} = 0 \qquad (6.4)$$

Hr, Hφ, Hz are the components of the magnetic field, H, along r, φ, and z axes, and $\vec{J}$ is current density. From equations (6.1) through (6.4) the following solutions are obtained for the calculation of the magnetic field of a current-carrying conductor.

Considering a simple case of an infinitely long cylindrical conductor carrying a direct current density, $\vec{J}$ (A/m²), as shown in Figure 6.1, we have,

$$\frac{\partial}{\partial \varphi} = 0 \text{ (due to circular symmetry)}$$

$$\frac{\partial}{\partial z} = 0 \text{ (due to infinitely long conductor)}$$

The current I through the conductor is,

$$I = J_z \pi R^2 \qquad (6.5)$$

where R is conductor radius.

Transmission Line Electric and Magnetic Fields

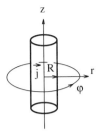

FIGURE 6.1 Magnetic field of a conductor carrying dc current j.

Field Outside of Conductor from (6.1)–(6.4)

$$\vec{j}_o = 0 \tag{6.6}$$

$$\nabla \times H = 0 \tag{6.7}$$

$$(\nabla \times H)_r = \frac{1}{r}\frac{\partial Hz}{\partial \varphi} - \frac{\partial H\varphi}{\partial z} = 0 \tag{6.8}$$

$$(\nabla \times H)_\varphi = \frac{1}{r}\frac{\partial Hr}{\partial z} - \frac{\partial Hz}{\partial r} = 0 \tag{6.9}$$

$$(\nabla \times H)_z = \frac{1}{r}H\varphi + \frac{\partial H\varphi}{\partial r} - \frac{\partial Hr}{\partial \varphi} = 0 \tag{6.10}$$

Since $\dfrac{\partial}{\partial z} = 0$

from (6.6) and (6.8) we obtain,

$$\frac{\partial Hz}{\partial r} = 0 \quad \text{or} \quad Hz = \text{constant} = 0$$

Since $\dfrac{\partial}{\partial \varphi} = 0$

from (6.10) we have,

$$\frac{1}{r}H\varphi + \frac{\partial H\varphi}{\partial r} = 0 \tag{6.11}$$

and obtain,

$$H\varphi = \frac{A}{r} \quad (6.12)$$

where A is a constant.
From (6.4) we have,

$$\nabla \cdot H = 0$$

$$\nabla \cdot H = \frac{1}{r} Hr + \frac{\partial Hr}{\partial r} + \frac{1}{r}\frac{\partial H\varphi}{\partial \varphi} + \frac{\partial Hz}{\partial z} = 0 \quad (6.13)$$

since,

$$\frac{\partial}{\partial \varphi} = 0, \quad \frac{\partial}{\partial z} = 0 \quad (6.14)$$

we have,

$$\frac{1}{r} Hr + \frac{\partial Hr}{\partial r} = 0 \quad (6.15)$$

and obtain the following solution,

$$Hr = \frac{B}{r} \quad (6.16)$$

where B is a constant.

Field Inside the Conductor

The current, $\vec{J}$, inside the conductor has the following components,

$J_r = 0$
$J_\varphi = 0$
$J_z \neq 0$

Since $\frac{\partial}{\partial \varphi} = 0$, from (6.10) we have,

$$\frac{1}{r} H\varphi + \frac{\partial H\varphi}{\partial r} = j_z \quad (6.17)$$

Transmission Line Electric and Magnetic Fields

which has for solution,

$$H\varphi = \frac{K}{r} + Lr \qquad (6.18)$$

where K and L are two constants which are determined from the following boundary conditions,

$$\text{as } r \to 0, \; H\varphi \to \infty$$

$$\text{or } K = 0$$

substituting, we have

$$\frac{\partial H\varphi}{\partial r} = -\frac{K}{r^2} + L$$

$$\frac{Lr}{r} + L = j_z \qquad (6.19)$$

and obtain,

$$H\varphi = \frac{I \cdot r}{2 \cdot \pi \cdot R^2} \qquad (6.20)$$

Summary of Equations

The magnetic intensity, $\vec{H}$ (A/m), inside and outside a conductor with current, I, is found as follows:

Inside the conductor: $\quad 0 \leq r \leq R$

$$H_r = 0 \qquad (6.21)$$

$$H_\phi = \frac{Ir}{2\pi \cdot R^2} \qquad (6.22)$$

$$H_z = 0 \qquad (6.23)$$

Outside the conductor: $\quad r > R$

$$H_r = 0 \qquad (6.24)$$

$$H_\phi = \frac{I}{2\pi \cdot r} \qquad (6.25)$$

$$H_z = 0 \qquad (6.26)$$

For environmental impact studies, we are interested in the field outside the conductor in free space. Equation (6.25) gives the magnetic field, H_ϕ, outside the conductor. The radial, H_r, and horizontal component, H_z, of the magnetic field outside the conductor is zero. It is seen that the magnetic field at ground level increases with transmission line ampacity (I) and by lowering of the distance (r) from conductor to ground. The distance (r) is a function of conductor sag.

6.2.2 THE MAGNETIC FIELD OF A THREE-PHASE POWERLINE

The magnetic field of a polyphase transmission line at a point in space can be calculated from (6.25) by vector addition of the magnetic field of each conductor as follows:

$$\vec{H}_r = \sum_{i=1}^{n} \vec{H}_i \qquad (6.27)$$

Therefore, for a three-phase transmission line having one conductor per phase, the magnetic field is,

$$\vec{H}_r = \vec{H}_1 = \vec{H}_2 + \vec{H}_3 \qquad (6.28)$$

$\vec{H}_r$ is the resultant magnetic field (A/m) of the transmission line at a point in space and $\vec{H}_1, \vec{H}_2, \vec{H}_3$ are the individual contributions of the magnetic field of each phase conductor at the same point in space.

Example 6.1

Calculate the magnetic field of a three-phase single circuit 750 kV transmission line at a point, M, 1 m above ground, and at a distance 30 m from the line shown in Figure 6.2. The line is loaded to its summer maximum thermal rating of 1500 A per phase.

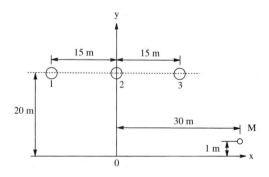

FIGURE 6.2 Phase configuration of 750 kV line (Example 6.1).

Solution
Transmission Line Configuration
Phase distance,

$x_1 = -15$ m
$x_2 = 0$ m
$x_3 = 15$ m
$y_1 = 20$ m
$y_2 = 20$ m
$y_3 = 20$ m

Phase current,

$\bar{I}_1 = 1500\angle 0$
$\bar{I}_2 = 1500\angle -120$
$\bar{I}_3 = 1500\angle 120$

The magnetic field at point M is calculated from (6.25),

$$\bar{H}_{i,m} = \frac{\bar{I}_i}{2\pi r_{i,m}}$$

I = phase current, A
i = 1,2,3 phase
$r_{im,}$ = distance from phase conductor i to point m, m

$$\bar{H}_{1,m} = \frac{1500\angle 0}{2\pi\sqrt{19^2 + 45^2}}$$

$$= \frac{1500\angle 0}{2\pi 48.8}\left[\frac{(20-1)}{48.8}\bar{u}_x + \frac{(15+30)}{48.8}\bar{u}_y\right]$$

$= 1.9\bar{u}_x + 4.5\bar{u}_y \; (\bar{u}_x, \bar{u}_y$ are unit vectors in x and y direction$)$

$$\bar{H}_{2,m} = \frac{1500\angle -120}{2\pi\sqrt{19^2 + 30^2}}$$

$$= \frac{1500\angle -120}{2\pi 35.5}\left[\frac{19}{35.5}\bar{u}_x + \frac{30}{35.5}\bar{u}_y\right]$$

$= (1.79 + j3.09)\bar{u}x + (2.84 + j4.88)\bar{u}y$

$$\overline{H}_{3,m} = \frac{1500\angle 120}{2\pi\sqrt{19^2 + 15^2}}$$

$$= \frac{1500\angle 120}{2\pi 35.5}\left[\frac{19}{24.2}\vec{u}x + \frac{15}{24.2}\vec{u}y\right]$$

$$= (3.87 - j6.65)\vec{u}x + (3.05 + j5.25)\vec{u}y$$

The magnetic field at point M is obtained by,

$$\vec{H}_M = \overline{H}_{1,m} + \overline{H}_{2,m} + \overline{H}_{3,m}$$

$$\boxed{\vec{H}_M = 5.38\angle 15.5 \quad A/m}$$

The magnetic field of a three phase powerline may be obtained approximately* by,

$$B = A \cdot \left(\frac{P^n}{r^{n+1}}\right) \cdot I \qquad (6.29)$$

$B = \mu_0 H$ is the magnetic field strength, Tesla
μ_0 = permeability of free space
H = magnetic field intensity, A/m
I = positive sequence current, A
P = phase to phase distance, m
r = distance from the axis of the line to the point of measurement, m
A = numerical coefficient depending upon line design (geometry)
n = number of sub-phases for a split phase line

The above equation is consistent with Equation (6.25) after taking into consideration the effect of multiple conductors, three-phase AC, and the transmission line geometry.

6.2.3 THE MAGNETIC FIELD OF DIFFERENT TRANSMISSION LINE GEOMETRY

In this section results are presented to show the magnetic fields of high-voltage transmission lines having different geometry. The calculations are based upon very optimistic line ampacities that would only be possible by adopting a dynamic line rating system. Conventional transmission line conductor configurations are shown in Figures 6.6, 6.7, and 6.8, and a new conductor configuration is shown in Figure 6.9. The geometry of Figure 6.9 provides a compact transmission line design with reduced magnetic field. The magnetic field of each configuration is shown in Figure 6.10, and the magnetic field at 30 m distance from the axis of the transmission line for each line configuration is shown in Table 6.1.

---

* V.S. Rashkes, R. Lordan, "Magnetic Field Reduction Methods: Efficiency and Costs," *IEEE Transactions on Power Delivery*, Vol. 13, No. 2, April 1998.

FIGURE 6.6 Single circuit horizontal configuration

FIGURE 6.7 Single circuit delta configuration

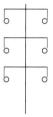

FIGURE 6.8 Double circuit vertical configuration

FIGURE 6.9 Compact line with phase splitting

The above values are well within the acceptable limit* of 1330 microTeslas for continuous exposure. A Cigré 1998 paper (Bohme et al. 1998) indicates 100 microTesla as the upper limit. It must also be mentioned that transmission line magnetic fields shown in Figure 6.10 are based upon high transmission line

---

* Restriction on human exposures to static and time-varying EM fields and radiation. Documents of the NRPB 4(5): 1–69, 1993. Exposure limits for power-frequency fields, as well as static fields and MW/RF frequencies; the standards apply to both residential and occupational exposure. For 60-Hz the limits recommended are 10 kV/m for the electric field and 1,330 micro T for the magnetic field. Copyright©, 1993–1998, by John E. Moulder, Ph.D. and the Medical College of Wisconsin.

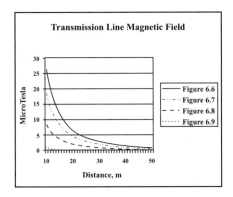

FIGURE 6.10 Transmission line magnetic field, ACSR Zebra 1000 A

TABLE 6.1

Configuration	Magnetic Field at 30 m, µT
Horizontal, single circuit, Figure 6.6	2.92
Delta, single circuit, Figure 6.7	2.06
Vertical, double circuit, Figure 6.8	0.94
Compact Star, phase splitting, Figure 6.9	0.01

conductor ampacity, 1000 A for an ACSR Zebra conductor. During normal operating conditions the line current will be less than 1000 A, and, therefore, the magnetic field will be lower.

6.2.4 EMF Mitigation

Even though overhead transmission lines are designed with low levels of EMF, even tighter control over the level of EMF radiated from a line can be achieved by EMF mitigation measures. Compact line designs having low levels of EMF are now used extensively.* Active and passive shielding of lines by the addition of auxiliary conductors on certain sections of the line are also used to lower the magnetic field at critical locations. These approaches have resulted in lowering the magnetic field to about 0.2 µT at the edge of transmission line right-of-way, and are recommended for areas such as schools, hospitals, and other areas where the public may be exposed to EMF continuously (Bohme et al., 1998).

Passive Shielding

Passive shielding of overhead lines is accomplished by the addition of auxiliary shield wires connected in a loop at certain critical sections of the line where

---

* "Compacting Overhead Transmission Lines," Cigré Symposium, Leningrad, USSR, 3–5 June, 1991.

controlling line EMF is important. In this method, current flows through the auxiliary conductor by induction from the powerline conductor. The magnitude and phase of the current in the auxiliary conductor is managed by controlling loop impedance. In this manner the auxiliary conductor generates a field that opposes the field produced by the power conductor, thereby lowering the total field from the line.

Active Shielding

Active shielding is similar to passive shielding, but instead of induced current in the auxiliary conductors, an external power supply is used to circulate a current through the shield wires. In this manner even greater control over the field generated by a line is possible by controlling the amplitude and phase of the current through the shield wires. The following example will help in the analysis of magnetic field shielding of overhead powerlines.

Example 6.2

Calculate the magnetic field of a three-phase 750 kV single circuit transmission line at a point M, 1 m above ground and 30 m from the line shown in Figure 6.2. The line is loaded to its summer maximum thermal rating of 1500 A per phase. Two auxiliary conductors, M_1 and M_2, shown in Figure 6.3, are used for magnetic field shielding by forming a 1 km current loop.

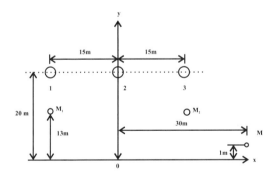

FIGURE 6.3 750 kV line with auxiliary shield wires (Example 6.2)

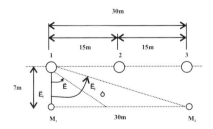

FIGURE 6.4 750 kV line showing shielding angles (Example 6.2)

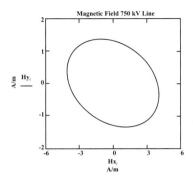

FIGURE 6.5 Magnetic field of 750 kV 3 phase transmission line with auxiliary shield wires.

Solution

The current I_1 flowing through the loop is calculated from,

$$\bar{I}_1 = \frac{\bar{V}_1}{\bar{Z}_1}$$

where,

$\bar{V}_1$ = voltage induced in the loop
$\bar{Z}_1$ = loop impedance

The induced voltage is calculated from

$$\bar{V}_1 = j\omega\phi\ell$$

where,

ω = angular frequency, radian/s
ϕ = total flux penetrating the loop, wb/m
ℓ = length of the loop

The total flux ϕ is calculated by

$$\phi = \int_S B.dS$$

where

S = area of the loop, m²
B = total flux density, Tesla

Since only the y component of the flux density vector $\vec{B}$ contributes to the above surface integral,

$$\vec{B} = \vec{B}_{1y} + \vec{B}_{2y} + \vec{B}_{3y}$$

$\vec{B}_{1y}, \vec{B}_{2y}, \vec{B}_{3y}$ are the flux density components due to current $\bar{I}_1, \bar{I}_2, \bar{I}_3$ in the phase conductors respectively and produce the flux $\bar{\phi}_1, \bar{\phi}_2, \bar{\phi}_3$, respectively.

Calculating $\bar{\phi}_1$ from the integral of equation,

$$\bar{\phi}_1 = \int_s \bar{B}_{1y} \, dn \cdot dz$$

Substituting

$$B = \mu_0 H$$

H from (6.25) we have,

$$\vec{B}_{1y} = \frac{\mu_0 \bar{I}_1 \sin\theta}{2\pi\rho}$$

from,

$$dx = h \cot(\text{cos}(\theta))$$

and,

$$dx = h \cot(\theta) d\theta$$

$$dz = 1$$

we find the flux $\bar{\phi}_1$ due to current $\bar{I}_1$,

$$\bar{\phi}_1 = \frac{\mu_0 \bar{I}_1}{2\pi} \int_{\theta_1}^{\theta_2} \sin(\theta)\cos(\theta)\cot(\theta) d\theta$$

From figure we have,

$$\theta_1 = 0$$

$$\theta_2 = 76.8^0$$

and obtain the solution to the integral,

$$\bar{\phi}_1 = 2.35 \cdot 10^{-4} \angle 0 \text{ wb}$$

Similarly, the flux $\bar{\phi}_2$ due to current $\bar{I}_2$ is calculated

$$\bar{I}_2 = 1500 \angle -120$$

$$\theta_1 = 64.6^0$$

$$\theta_2 = 64.6^0$$

$$\bar{\phi}_2 = \frac{\mu_0 \bar{I}_2}{2\pi} \int_{\theta_1 = 64.6}^{\theta_2 = 64.6} \sin(\theta)\cos(\theta)\cot(\theta)d\theta$$

$$\bar{\phi}_2 = 4.55 \cdot 10^{-4} \angle -120 \text{ wb}$$

Flux ϕ_3 due to current I_3,

$$\bar{I}_3 = 1500 \angle 120$$

$$\theta_1 = -76.6^0$$

$$\theta_2 = 0^0$$

$$\bar{\phi}_3 = \frac{\mu_0 \bar{I}_3}{2\pi} \int_{\theta_1 = 76.6}^{\theta_2 = 0} \sin(\theta)\cos(\theta)\cot(\theta)d\theta$$

$$\bar{\phi}_3 = 2.35 \cdot 10^{-4} \angle 120 \text{ wb}$$

The total flux is,

$$\bar{\phi} = \bar{\phi}_1 + \bar{\phi}_2 + \bar{\phi}_3$$

$$\bar{\phi} = 2.2 \cdot 10^{-4} \angle -120 \text{ wb}$$

The induced loop voltage $\bar{V}_\ell$ is,

$$\bar{V}_\ell = j \cdot \phi \cdot \omega \cdot \ell$$

$\ell = 1000$ m (1 km loop)
$\omega = 2\cdot\pi\cdot 60$ rad/s
$\bar{V}_\ell = j(2.2 \cdot 10^{-4} \angle -120) \cdot 377 \cdot 10^3$
$\bar{V}_\ell = 83.2 \angle -30$

Transmission Line Electric and Magnetic Fields

The auxiliary conductor loop impedance $\overline{Z}a$ is selected as,

$$\overline{Z}a = 0.3 \angle -30, \ \Omega$$

The loop current is,

$$\overline{I}_\ell = \frac{83.2 \angle -30}{0.3 \angle -30}$$

$$\overline{I}_\ell = 277.3 \ A$$

Magnetic Field with Shielding

The magnetic field, Ha, at point M, due to the two auxiliary conductors, is calculated, as in Example 6.1, from,

$$\overline{H}a_{i,m} = \frac{Ia_i}{2\pi r_{i,m}}$$

Ia = current in auxiliary conductors, A
i = 1,2
$r_{i,m}$ = distance from auxiliary conductors, conductor i to point M

$$\overline{H}a_{1,m} = \frac{277.3}{2\pi\sqrt{12^2 + 45^2}}$$

$$= \frac{277.3}{2\pi 46.57}\left[\frac{(13-1)}{46.57}\vec{u}_x + \frac{(15+30)}{46.57}\vec{u}_y\right]$$

$$= -0.24\vec{u}_x - 0.92\vec{u}_y$$

$$\overline{H}_{2,m} = \frac{277.3}{2\pi\sqrt{12^2 + 45^2}}$$

$$= \frac{277.3}{2\pi 19.2}\left[\frac{12}{19.2}\vec{u}x + \frac{15}{19.2}\vec{u}y\right]$$

$$= -1.43\vec{u}x - 1.79\vec{u}y$$

$$\vec{H}_{a,x} = -(0.24 + 1.43)\vec{u}x = -1.67\vec{u}x$$

$$\vec{H}_{a,y} = -(0.92 + 1.79)\vec{u}y = -2.71\vec{u}y$$

$$\vec{H}_x = \vec{H}_{M,x} + \vec{H}_{a,x} = \{(3.77 - j3.56) - 1.67\}\bar{u}x = (2.08 - j3.56)\bar{u}x$$

$$\vec{H}_y = \vec{H}_{M,y} + \vec{H}_{a,y} = \{(1.93 - j0.37) - 2.71\}\bar{u}y = (1.31 - j0.37)\bar{u}y$$

The magnetic field $\vec{H}$ is an ellipse

$$\left|\vec{H}\right| = \sqrt{\vec{H}_x^2 + \vec{H}_y^2} = 4.3$$

$$\text{Angle} = \tan^{-1}\left(\frac{|H_y|}{|H_x|}\right) = 18.3^0$$

$$\vec{H} = 4.3 \angle 18.3$$

6.3 TRANSMISSION LINE ELECTRIC FIELD

The electric field, E, of the transmission line at any point in space is a function of line voltage and the distance of the point from the transmission line conductor. Therefore, the electric field at ground level is affected by conductor sag since an increase in conductor temperature will increase sag and result in lowering the distance of the powerline conductor to the ground. The effect of conductor sag due to higher transmission line ampacity was studied in Chapter 5. In this section we shall study the method of calculation of the electric field of a transmission line and determine the effect of conductor sag on the electric field at ground level.

The electric field strength E may be defined as gradient of the potential V given as,

$$\vec{E} = -\text{grad}(\vec{V}) \text{ V/m} \qquad (6.30)$$

There exists an electric field if there is a potential difference between two points having potential V_1 and V_2 separated by a distance, r, such that,

$$\text{grad}(\vec{V}) = \frac{\vec{V}_1 - \vec{V}_2}{r} \qquad (6.31)$$

In a perfect conductor the potential difference ($\vec{V}_1 - \vec{V}_2$) is zero, the gradient of the voltage is effectively zero, and, hence, the electric field inside a perfect conductor is also zero.

Electric field calculation

The various methods of calculation of the electric field by numerical and analytical methods are given in a Cigré report (Cigré, 1980). In this section we shall apply the

Transmission Line Electric and Magnetic Fields

analytical method of equivalent charges to calculate the electric field of three-phase transmission lines.

The electric field, E, at a distance, r, from a charge, q, is calculated by Gauss's law,

$$E = \frac{q}{2\pi\varepsilon_0 r} \qquad (6.32)$$

The q charges carried by transmission line conductors is calculated by,

$$[q] = [C] \cdot [V] \qquad (6.33)$$

where [q] is a column vector of charges, [C] is the capacitance matrix of the multi-conductor circuit, and [V] a column vector of phase voltages.

The capacitance matrix [C] is calculated from Maxwell's potential coefficients [λ] defined as the ratio of the voltage to charge.

The elements, λ_{ii}, of the matrix of potential coefficients are calculated by,

$$\lambda_{ii} = \frac{1}{2\pi\varepsilon_0} \ln \frac{2h_i}{r_i} \qquad (6.34)$$

where,

h_i = height of the conductor i above ground
r_i = radius of conductor i

For bundle conductor system an equivalent radius is calculated as,

$$r_{eq} = R \cdot n \sqrt{\frac{nr}{R}} \qquad (6.35)$$

r = subconductor radius
n = number of subconductors in bundle
R = geometric radius of the bundle

The elements λ_{ij} of the matrix of potential coefficients are calculated by,

$$\lambda_{ij} = \frac{1}{2\pi\varepsilon_0} \ln \frac{2D'_{ij}}{D_{ij}} \qquad (6.36)$$

where $D_{i,j}$ is the distance between conductors i and j, and $D'_{i,j}$ is the distance between image conductors i' and j'.

The matrix [C] is calculated by inversion,

$$[C] = [\lambda]^{-1} \tag{6.37}$$

Knowing [C] and [V], we calculate [q] from,

$$[q] = [C] \cdot [V] \tag{6.38}$$

The electric field, E, is then calculated by the application of Gauss's law by vector summation of the individual fields due to the charge on each conductor

$$\vec{E} = \sum_{i=1}^{i=n} \vec{E}_i \tag{6.39}$$

where,

$$\vec{E}_i = \frac{q}{2\pi\varepsilon_0 r_i} \tag{6.40}$$

The following example illustrates the important concepts presented in this section by showing the calculation of the electric field of the transmission line in Example 1.

Example 6.3

Calculate the electric field of a three phase single circuit 750 kV transmission line at a point, M, 1 m above ground, and at a distance 30 m away from the line as shown in Figure 6.1. The line is loaded to its summer maximum thermal rating of 1500 A per phase.

FIGURE 6.6 Three-phase configuration of a 750 kV transmission line

Solution

The height of conductor above ground is given as,

$$h_1 = h_2 = h_3 = 20\text{m}$$

The equivalent radius of four conductor bundle is,

$$r_{eq} = R \cdot n\sqrt{\frac{nr}{R}}$$

$$r_{eq} = 150 \cdot 10^{-3} \cdot 4\sqrt{\frac{4 \cdot 30 \cdot 10^{-3}}{150 \cdot 10^{-3}}}$$

$$= 0.54 \text{ m}$$

Maxwell's potential coefficients are easily calculated in Mathcad® as,

$$\lambda(n, m) := \begin{vmatrix} \text{for } i \in 1..n \\ \quad \text{for } j \in 1..m \\ \quad\quad A_{i,j} \leftarrow \dfrac{\ln\left(2 \cdot \dfrac{h_i}{r_i}\right)}{2 \cdot \pi \cdot \varepsilon 0} \text{ if } i = j \\ \quad\quad A_{i,j} \leftarrow \dfrac{\ln\left[\dfrac{(D'_{i,j})}{D_{i,j}}\right]}{2 \cdot \pi \cdot \varepsilon 0} \text{ otherwise} \\ A \end{vmatrix}$$

For a three-phase transmission line, n = m = 3, resulting in the following matrix of potential coefficients,

$$\lambda(3,3) = \begin{bmatrix} 7.76 \cdot 10^{10} & 1.884 \cdot 10^{10} & 9.195 \cdot 10^{9} \\ 1.884 \cdot 10^{10} & 7.76 \cdot 10^{10} & 1.884 \cdot 10^{10} \\ 9.195 \cdot 10^{9} & 1.884 \cdot 10^{10} & 7.76 \cdot 10^{10} \end{bmatrix}$$

And we obtain the capacitance matrix [C] from,

$$[C] = [\lambda]^{-1}$$

$$C = \begin{bmatrix} 1.375 \cdot 10^{-11} & -3.126 \cdot 10^{-12} & -8.7 \cdot 10^{-13} \\ -3.126 \cdot 10^{-12} & 1.44 \cdot 10^{-11} & -3.126 \cdot 10^{-12} \\ -8.7 \cdot 10^{-13} & -3.126 \cdot 10^{-12} & 1.375 \cdot 10^{-11} \end{bmatrix}$$

The charge, q, is obtained by substitution of [C] and [V],

$$[q] = [C][V]$$

$$[q] = \begin{bmatrix} 1.375 \cdot 10^{-11} & -3.126 \cdot 10^{-12} & -8.7 \cdot 10^{-13} \\ -3.126 \cdot 10^{-12} & 1.44 \cdot 10^{-11} & -3.126 \cdot 10^{-12} \\ -8.7 \cdot 10^{-13} & -3.126 \cdot 10^{-12} & 1.375 \cdot 10^{-11} \end{bmatrix} \begin{bmatrix} 750 \cdot 10^3 \angle 0 \\ 750 \cdot 10^3 \angle 120 \\ 750 \cdot 10^3 \angle 240 \end{bmatrix}$$

giving,

$$[q] = \begin{bmatrix} 1.18 \cdot 10^{-5} - j1.45 \cdot 10^{-6} \\ -6.574 \cdot 10^{-6} + j1.131 \cdot 10^{-5} \\ -4.636 \cdot 10^{-6} - j1.088 \cdot 10^{-5} \end{bmatrix}$$

The electric field, E, is obtained by the application of Gauss's law,

$$[E] = \frac{1}{2\pi\varepsilon_0} \begin{bmatrix} \frac{q_1}{\rho_1} \\ \frac{q_2}{\rho_2} \\ \frac{q_3}{\rho_3} \end{bmatrix}$$

The resultant E field at M is calculated by vector summation by adding X and Y components of the individual elements of [E]. The X and Y components are obtained as,

$$Ey_i := E_i \cdot \rho_i \cdot \left[\frac{(y_i - h_i)}{(x_i - d_i)^2 + (y_i - h_i)^2} - \frac{(y_i + h_i)}{(x_i - d_i)^2 + (y_i + h_i)^2} \right]$$

$$Ex_i := E_i \cdot \rho_i \cdot \left[\frac{(x_i - d_i)}{(x_i - d_i)^2 + (y_i - h_i)^2} - \frac{(x_i + d_i)}{(x_i - d_i)^2 + (y_i + h_i)^2} \right]$$

Transmission Line Electric and Magnetic Fields

$$\sum_{i=1}^{3} Ey_i = 9.475 \cdot 10^3 + 7.602i \cdot 10^3$$

$$\sum_{i=1}^{3} Ex_i = -127.271 + 1.181i10^3$$

The resultant electric field is an ellipse as seen in the Figure 6.7.

FIGURE 6.7 The electric field of 750 kV three-phase transmission line.

6.4 CHAPTER SUMMARY

The magnetic field of a transmission line at ground level is a function of line current and the distance of phase conductors from ground. The magnetic field inside and outside of a current-carrying conductor has been developed from Maxwell's equation and Ampere's law. As shown by the equations developed in this chapter, an increase in line current increases the magnetic field at ground level. The magnetic field at ground level also increases with higher sag. If conductor temperature is higher than normal due to higher current through the line, then the magnetic field at ground level will become more significant due to the combined effect of high current and reduced distance of conductor to ground. A numerical example is provided in this chapter for the calculation of the magnetic field of three-phase transmission line. Methods of reducing magnetic fields by active and passive shielding are also presented in this chapter.

The electric field from a transmission line at ground level is indirectly affected by line ampacity only if an increase in line ampacity raises the maximum design temperature of the transmission line conductor. If conductor temperature is higher than the maximum allowed for the line, then sag will increase. Consequently the distance from conductor to ground will become less than normal which will raise the electric field at ground level. A method of calculation of electric field at ground level due to higher transmission line ampacity is given in this chapter with a numerical example.

Since both electric and magnetic fields of a transmission line depend upon conductor temperature, it is very important that increasing line ampacity does not exceed the maximum design temperature of the conductor. Therefore, for EMF considerations also, it is important to follow a dynamic line rating system that will maintain normal conductor temperature within a specified limit.

7 Weather Modeling for Forecasting Transmission Line Ampacity

7.1 INTRODUCTION

Since weather is an important parameter in the determination of transmission line ampacity, the development of weather models of ambient temperature, wind speed, wind direction, and solar radiation are presented in this chapter. These are statistical weather models based upon time-series analysis and National Weather Service forecasts. Hourly values of future meteorological conditions from 1 to 24 hours ahead, or even up to 1 week in advance, are now becoming possible due to developments in weather forecasting.

The solution of the differential equations for the heating of a conductor by current in the steady, dynamic, and transient states requires the knowledge of the following meteorological variables:

- Ambient Temperature
- Wind Speed
- Wind Direction
- Solar Radiation

When transmission line ampacity is required for the present time, the above meteorological data can be obtained by measurement from weather stations. For the prediction of line ampacity several hours in advance, a weather model is required. In this chapter, stochastic and deterministic models of ambient temperature, wind speed, and wind direction, and an analytical model of solar radiation shall be developed from time series data. Neural network models are also presented for forecasting hourly values of meteorological conditions as well as for weather pattern recognition.

The prediction of transmission line ampacity several hours in advance has become more important today due to competition in the electric power supply industry, and greater need for the advance planning of electricity generation and transmission capacity* (Deb, 1998, 1997, 1995; Cibulka, Williams, and Deb, 1991; Hall and Deb, 1988c; Douglass, 1986; Foss and Maraio, 1989). Numerical examples

---

* M. Aganagic, K. H. Abdul-Rahman, J.G. Waight. *Spot pricing of capacities for generation and transmission of reserve in an extended Poolco model.* IEEE Transactions on Power Systems, Vol. 13, No. 3, August 1998

of weather forecasting are presented in this chapter, followed by an example of line ampacity forecast generated by a program.

Having precise knowledge of future transmission line capacity will greatly facilitate the purchase of competitively priced electricity from remote locations. In the future we expect an increase in the number of power producers requiring access to utility transmission lines for the supply of electricity. For these reasons, a transmission line ampacity program with forecast capability is essential.

The LINEAMPS program uses a weather model based on historical weather data as well as weather forecast data prepared by the National Weather Service. Two alternative approaches to weather modeling are developed. In the first approach, hourly values of historical weather data for different seasons of the year are fitted by a Fourier series. In the second approach, weather patterns are recognized by training an unsupervised neural network using Kohonen's learning algorithm (Haykin, 1999; Eberhart and Dobbins, 1990). These patterns are then adjusted to forecast weather data available from the National Weather Service or other weather service companies. Hourly values of future ambient temperature and wind speed data are then generated from these patterns as described in the following section.

When continuous input of real-time meteorological data is available, a Kalman filter-type algorithm is developed for the recursive estimation of weather variables for real-time prediction of transmission line ampacity.

7.2 FOURIER SERIES MODEL

Hourly values of ambient temperature (Ta) and wind speed (W_s) at time (t) are generated by AmbientGen and WindGen methods in the weather station object of the program by fitting Fourier series to historical weather data of the region. The Fourier series model is given by,

$$Y(t) = A_0 + k \sum_{i=1}^{n} C_i \mathrm{Sin}(\omega_i t) + \sum_{i=1}^{n} B_i \mathrm{Cos}(\omega_i t) \tag{7.1}$$

Description of symbols

$Y(t) \in \{Ta(t), W_s(t)\}$
A_0, C_i, B_i for i = 1...n are coefficients of the model
$\omega = 2\pi/24$ = fundamental frequency
k = factor used to adjust Fourier series to National Weather Service forecast.
   It is calculated by,

$$k = \frac{Y_f(t_{max}) - Y_f(t_{min})}{F(t_{max}) - F(t_{min})} \tag{7.2}$$

$Y_f(t_{max})$ = daily maximum value of ambient temperature or wind speed forecast by the National Weather Service.

$Y_f(t_{min})$ = daily minimum value of ambient temperature or wind speed forecast by the National Weather Service.

$F(t_{max})$ and $F(t_{min})$ are found from the Fourier series (7.1), when $t = t_{max}$ and $t = t_{min}$ respectively.

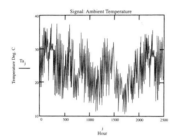

FIGURE 7.1 Hourly averaged ambient temperature data during summer time in San Francisco.

The development of a weather model (Figure 7.1) requires determining the parameters of the model from historical weather data of each of the meteorological variables. The unknown parameters of the model $[A_0, K, A_i, B_i, n, \omega]$ are determined by least square regression analysis. The fundamental frequency ω and the coefficients $r_N(\omega)$ of the frequency spectra are also determined by spectral analysis (Priestley 1981) as shown in the Figures 7.2, 7.4, and 7.6 where,

$$r_N(\omega) = \frac{2}{N} \left| \sum_{t=1}^{N} X_i \cdot e^{-j\omega t} \right| \qquad (7.3)$$

$$e^{-j\omega t} = A(\omega) + jB(\omega) \qquad (7.4)$$

From Figure 7.2 we see that the dominant frequency is equal to 0.042, which is also the fundamental frequency. Since the period $T = 1/f$, the fundamental period is found to be approximately equal to 24 hours, as we should expect for the region of San Francisco. The same phenomenon is observed in all of the meteorological variables comprising ambient temperature, wind speed, wind direction, and solar radiation in this region as seen in Figures 7.2, 7.4, and 7.6.

Examples of Fourier series patterns of hourly ambient temperature and wind speed that were developed for the San Francisco Bay area during summer time are shown in Figure 7.14. It is appropriate to mention here that these patterns are applicable to the region of the San Francisco Bay area only. A similar analysis is required for transmission lines in other regions.

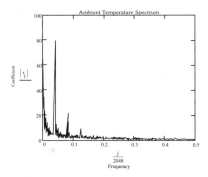

FIGURE 7.2 Ambient temperature spectrum.

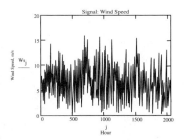

FIGURE 7.3 Hourly averaged wind speed data during summer time in San Francisco.

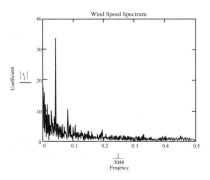

FIGURE 7.4 Wind speed spectrum.

The daily cyclical behavior of the meteorological variables is further supported by the autocorrelations that were calculated from the hourly averaged values of each time series as shown in the Figures 7.15-7.17.

The autocorrelation r_k is calculated by,

Weather Modeling for Forecasting Transmission Line Ampacity 119

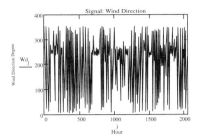

FIGURE 7.5 Hourly averaged wind direction data during summer time in San Francisco.

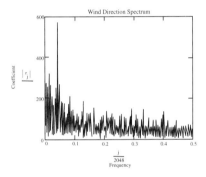

FIGURE 7.6 Wind direction spectrum.

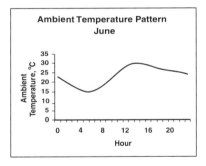

FIGURE 7.7 Ambient temperature pattern, June.

$$r_k = \frac{c_k}{c_0} \tag{7.5}$$

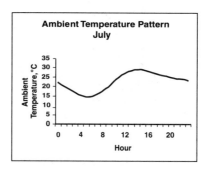

FIGURE 7.8 Ambient temperature pattern, July.

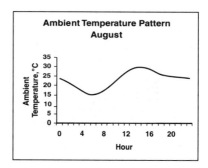

FIGURE 7.9 Ambient temperature pattern, August.

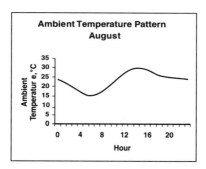

FIGURE 7.10 Ambient temperature pattern, September.

The autocovariance c_k is given by,

$$c_k = \frac{1}{N}\sum_{t=1}^{N-k}(z_t - \bar{z})(z_{t+k} - \bar{z}) \quad k = 0, 1, 2 \ldots n \text{ lags} \quad (7.6)$$

z_t = average hourly value of the meteorological variable at time t

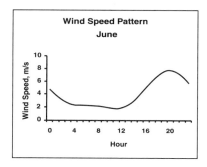

FIGURE 7.11 Wind speed pattern, June.

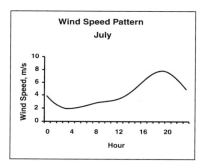

FIGURE 7.12 Wind speed pattern, July.

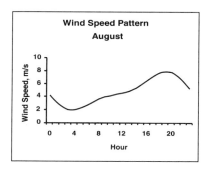

FIGURE 7.13 Wind speed pattern, August.

The process mean is,

$$\bar{z} = \frac{\sum_{t=1}^{N} z_t}{N} \quad (7.7)$$

N = number of observations in the time series

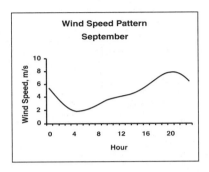

FIGURE 7.14 Wind speed pattern, September.

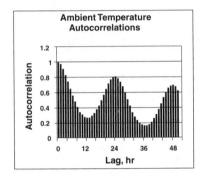

FIGURE 7.15 Ambient temperature autocorrelations.

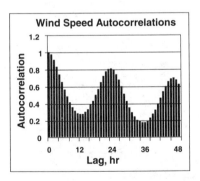

FIGURE 7.16 Wind speed autocorrelations.

One of the important advantages of a Fourier series model (Figure 7.1) for the prediction of hourly values of ambient temperature and wind speed is that it does not require continuous input of weather data. Hourly values of future weather data are generated by the model by adjusting the coefficients with general purpose weather forecast data. The daily maximum and minimum values forecast by the weather

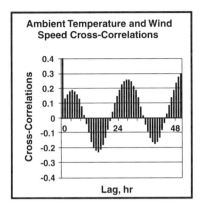

FIGURE 7.17 Ambient temperature and wind speed cross-correlations.

service are used to adjust model coefficients using Figure 7.2. Figure 7.19 shows hourly values of ambient temperature generated by the Fourier series model in comparison with measured data for one week during summer in the San Francisco Bay area. The model is also useful for simulation purpose as shown in Figure 7.20, and as discussed in Chapter 5.

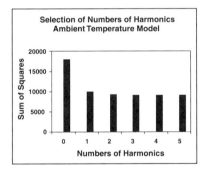

FIGURE 7.18 Selection of number of harmonics in the Fourier series model of ambient temperature.

7.3 REAL-TIME FORECASTING

As stated earlier, forecasting of transmission line ampacity several hours in advance is beneficial for the advance planning of generation and transmission resources. Due to deregulation in the electric utility industry there is even greater competition for the supply of electric energy. Utilities and power producers are making advance arrangements for the purchase and sale of electricity, which requires ensuring adequate transmission capacity. Transmission line capacities are predicted in advance by the LINEAMPS program by taking into account weather forecast data and the weather models developed in the previous section.

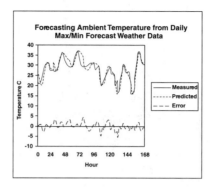

FIGURE 7.19 Forecasting hourly values of ambient temperature.

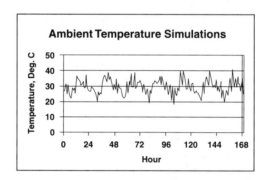

FIGURE 7.20 Simulation of hourly values of ambient temperature by Fourier series model and a second order autoregressive stochastic model.

7.3.1 FORECASTING AMPACITY FROM WEATHER PATTERNS

The algorithm for forecasting line ampacity by Fourier series weather patterns is given in the flow chart of Figure 7.31. An example of forecasting hourly values of line ampacity up to seven days in advance by the LINEAMPS program is given in the Figure 7.32.

7.3.2 REAL-TIME FORECASTING OF TRANSMISSION LINE AMPACITY

When real-time weather data is available continuously, it is possible to forecast hourly values of weather data on an hour-by-hour basis by the application of the Kalman filter algorithm, which is suitable for real-time predictions. It is a recursive algorithm that calculates future values of the meteorological variables consisting of ambient temperature, wind speed, and wind direction based on previous measurements of these variables. The predicted values of meteorological variables are then entered into the transmission line heat balance equation to calculate line ampacity. The following recursive algorithm is developed for the prediction of hourly values of ambient temperature, wind speed, and wind direction.

The measurements, y(t), comprising ambient temperature, wind speed, and wind direction are considered to be composed of a periodic component, p(t), and a stochastic component, z(t),

$$y(t) = p(t) + z(t) \tag{7.8}$$

where y(t) represents the hourly values of measurement of ambient temperature, wind speed, and wind direction, or a coefficient of heat transfer, Hc.

The Auto Regressive Moving Average process with exogenous Variables (ARMAV) was selected to model z(t) as given below,

$$A(q^{-1},t)z(t) = B(q^{-1},t)U(t-d) + C(q^{-1},t)S(t) \tag{7.9}$$

where, $A(q^{-1},t)$, $B(q^{-1},t)$, $C(q^{-1},t)$ are time-variable polynomials in the backward shift operator q^{-1}:

$$A(q^{-1},t) = 1 + a_1(t) q^{-1} + a_2(t) q^{-1} + \ldots + a_{na}(t) q^{-1}$$

$$B(q^{-1},t) = 1 + b_1(t) q^{-1} + b_2(t) q^{-1} + \ldots + b_{nb}(t) q^{-1}$$

$$C(q^{-1},t) = 1 + b_1(t) q^{-1} + b_2(t) q^{-1} + \ldots + c_{nb}(t) q^{-1}$$

In the above equations, z(t), u(t), and s(t) represent the output, input, and white noise sequence, respectively.

The periodic term, p(t), is represented by a Fourier series given by,

$$p(t) = m(t) + f_i(t) \sin(i\omega t) + g_i(t) \cos(i\omega t) \tag{7.10}$$

where,

m(t) = process mean
f_i, g_i i = 1,2 ... nh are the coefficients of the model
nh = number of harmonics
$\omega = 2\pi/24$ = fundamental period

Writing the parameter vector compactly,

$$x^T(t) = \{a_1(t) \ldots a_{na}(t), b_1(t) \ldots b_{nb}(t), c_1(t) \ldots c_{np}(t),$$

$$m(t), f_i(t) \ldots f_{nh}(t), g_i(t) \ldots g_{nh}(t)\} \tag{7.11}$$

the problem now becomes that of estimating $x^T(t)$ at each instant (t) based on the measurement y(t). This is carried out recursively by the Kalman filter algorithm.

State Equation

The true value of the parameter vector is assumed to vary according to,

$$x(t+1) = x(t) + v(t) \qquad (7.12)$$

where $v(t)$ is a sequence of independent gaussien random vector.

Measurement Equation

From equations (7.7) – (7.11) we may write the measurement equation as,

$$y(t) = H^T(t) x(t) + e(t) \qquad (7.13)$$

where the matrix H (actually a row vector) is given by,

$$\begin{aligned}H^T = \{ & -y(t-1), -y(t-2), \ldots -y(t-np), \\ & 1, u(t-d-1), u(t-d-2), \ldots u(t-d-nw), \\ & n(t-1), n(t-2), \ldots .n(t-nf), \\ & 1, \sin(\omega t), \sin(2\omega t), \ldots \sin(nh\omega t), \\ & \cos(\omega t), \cos(2\omega t), \ldots \cos(nh\omega t)\} \end{aligned} \qquad (7.14)$$

Equations (69), (70) constitute the state and measurement equation, respectively, and, therefore, the problem of parameter estimation is reduced to the problem of state estimation. The Kalman filter algorithm can now be applied to estimate the state vector $x(t)$.

7.3.3 KALMAN FILTER ALGORITHM

State Update Equation

$$x(t) = x(t-1) + n(t)k(t) \qquad (7.15)$$

Innovations

$$n(t) = y(t) - H^T x(t-1) \qquad (7.16)$$

Kalman Gain

$$k(t) = p(t-1)H(t)[R_2(t) + H^T(t)p(t-1)H(t)]^{-1} \qquad (7.17)$$

Weather Modeling for Forecasting Transmission Line Ampacity

where the error covariance matrix p(t) is given by,

$$p(t) = p(t-1) + \frac{\left[R_1(t) - p(t-1)H^T(t)p(t-1)\right]}{\left[R_2(t) + H^T(t)p(t-1)H(t)\right]} \quad (7.18)$$

The results obtained by the application of the above algorithm* to predict hourly values of ambient temperature are shown in the Figure 7.21.

FIGURE 7.21 Recursive estimation of San Francisco Bay area ambient temperature during summer time by the application of Kalman filter algorithm.

7.4 ARTIFICIAL NEURAL NETWORK MODEL

Neural network is an important subject of research in artificial intelligence where computations are based on mimicking the functions of a human brain. Neural networks consist of many simple elements called neurons that are linked by connections of varying strengths as shown in Figure 7.22. The neural networks used in numerical analysis today are gross abstractions of the human brain. The brain consists of very large numbers of far more complex neurons that are interconnected with far more complex and structured couplings (Haykin. 1999).

A supervised neural network using the back propagation algorithm, and an unsupervised neural network using Kohonen's learning algorithm, are the two types of neural networks that were used in weather modeling for line ampacity predictions. A neural network is trained to forecast hourly values of ambient temperature and wind speed by using the back propagation algorithm. An unsupervised neural network is also developed for weather pattern recognition by using Kohonen's learning algorithm. Results obtained by the application of the above neural networks are presented in Figures 7.23 and 7.24.

---

* There is recent interest in the application of the Kalman filter algorithm for the efficient solution of nonlinear recurrent neural networks for real-time prediction. For example, a recent book by Simon Haykin, *Neural Networks: A Comprehensive Foundation*, published by Prentice-Hall in 1999 recommends the Kalman filter for real-time recurrent learning.

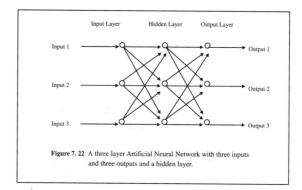

FIGURE 7.22 A three-layer artificial neural network with three inputs and three outputs and a hidden layer.

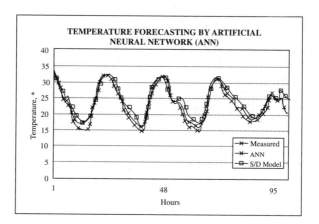

FIGURE 7.23 Example of neural network application to forecast next hour ambient temperature in the San Francisco Bay area. The network trains by supervised learning using the back propagation algorithm. Neural network results are compared to a statistical forecasting model and actual data.

According to Haykin,* a neural network is a massively parallel distributed processor that has a natural propensity for storing experiential knowledge and making it available for use. It resembles the brain in two respects:

1. Knowledge is acquired by the network through a learning process.
2. Interneuron connection strengths known as synaptic weights are used to store the knowledge.

---

* Haykin, S. (1994), Neural Networks: A Comprehensive Foundation, Macmillan, NY, p. 2.

Weather Modeling for Forecasting Transmission Line Ampacity 129

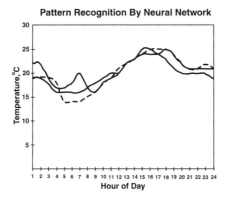

FIGURE 7.24 Application of unsupervised neural network for ambient temperature pattern recognition in the San Francisco Bay area.

7.4.1 Training of the Neural Network

A neural network is first trained by feeding it with data, from which it learns the input-output relationship of a system. Once trained, the network provides the correct output from a set of input data. If the training set is sufficiently large, then the neural net will provide the correct output from a set of input data that is different from the training set. A neural network is different from a look-up table. Unlike a look-up table, the dynamics of the system are represented by a trained neural network. It is therefore clear that a neural network is particularly useful to predict the outcome of a phenomenon that cannot be formulated otherwise.

7.4.2 Supervised and Unsupervised Learning

Learning is the key to AI—the artificial neural network learns from data and demonstrates intelligent capability. Learning in an artificial neural network is either supervised or unsupervised.

Supervised Learning

In a supervised neural network, learning or training is carried out by a back propagation algorithm, where network output is compared to a target and the difference is used to adjust the strength of the connections. Training is completed when the sum of squares of errors is minimized. The back propagation algorithm is presented below, and an application of this algorithm to forecast hourly ambient air temperature in the San Francisco Bay area is shown in Figure 7.23.

Unsupervised Learning

In unsupervised learning there is no teacher, in other words, there is no target response with which to compare output. The network organizes by itself (self-organizing neural network) and learns to recognize patterns within data. Unsupervised learning is carried out by Kohonen's learning algorithm, which is used for pattern recognition and data classification. In Kohonen's learning algorithm the strength of interconnections is adjusted by minimizing the Euclidean distance of output neuron. The output neuron having the minimum Euclidean distance is declared the winner and set to 1; all others are set to 0.

For the above reasons, a self-organizing neural network is also called a "winner-take-all" algorithm, because when the network is trained only a certain output will go high, depending upon the characteristics of the input vector. During training, the weights of the connections are adjusted until subsequent iterations do not change weights. A winner-take-all self-organizing neural network due to Kohonen's learning algorithm is presented below, and its application to pattern recognition of daily ambient temperature is shown in Figure 7.24. The above types of neural networks are examples of nonrecurrent networks.

Another important type of network is the recurrent network due to Hopfield, where there is continuous feedback from output to input. Recurrent networks find applications in nonlinear optimization problems whose solutions are difficult by conventional means.

7.4.3 BACK PROPAGATION ALGORITHM*

The back propagation algorithm is composed of the following steps:

1. Apply an input vector **x**
2. Calculate the error **e** between the output vector **y** and a known target vector **z**

$$e = \sum_{l=1}^{n}\left[z(l) - y(l)\right]^2 \quad n = \text{length of training vector}$$

3. Minimize errors

$$\frac{\partial e}{\partial w(j, i)} = \frac{\partial e}{\partial y} \cdot \frac{\partial z}{\partial w(l, j)} = \delta(l)$$

4. Calculate error signal of output layer $\delta(l)$

$$\delta(l) = [z(l) - y(l)] \cdot y(l) \cdot [1 - y(l)]$$

5. Calculate error signal of input layer $\delta(j)$

$$\delta(j) = y(j) \cdot [1 - y(j)] \sum_{l=0}^{l=nl} w(l, j)\delta(l)$$

---

* Rumelhart, David E., McClelland, James, L., *Parallel Distributed Processing*, Volume 1, M.I.T. Press, Cambridge, MA, 1988.

6. Update weights by the learning rule

$$w(j, i) \text{ new} = w(j, i) \text{ old} + \delta(j) \cdot y(i) + \alpha[\Delta w(j, i) \text{old}]$$

$$w(l, i) \text{ new} = w(l, i) \text{ old} + \delta(l) \cdot y(l) + \alpha[\Delta w(l, i) \text{old}]$$

Steps 1 through 6 are repeated until the sum of squares of errors is minimized.

A neural network for the prediction of hourly values of ambient temperature is developed by using the back propagation algorithm and the results are presented in Figure 7.23.

7.4.4 Unsupervised Neural Network Training Algorithm*

The unsupervised neural network training algorithm is due to Kohonen and is composed of the following steps:

1. Apply an input vector **x**
2. Calculate the Euclidean space **d(j)** between **x** and the weight vector **w** of each neuron as follows:

$$d(j) = \sqrt{\sum_{i=1}^{n} [(x(i) - w(i, j))]^2}$$

n = number of training vector x
$w(i, j)$ = weight from input i to neuron j

3. The neuron that has the weight vector closest to **x** is declared the winner. This weight vector, called w_c, becomes the center of a group of weight vectors that lie within a distance **d** from w_c.
4. Update nearby weight vectors as follows:

$$w(i, j) \text{ new} = w(i, j) \text{ old} + \alpha[x - w(i, j) \text{ old}]$$

where, α is a time-varying learning coefficient normally in the range $0.1 < \alpha < 1$. It starts with a low value of 0.1 and gradually increases to 1 as learning takes place. Steps 1 through 4 are repeated until weight change between subsequent iterations becomes negligible.

A self-organizing neural network is also developed for ambient temperature pattern recognition by using Kohonen's learning algorithm, and the results are presented in Figure 7.25.

---

* Wasserman, Philip D. 1989 *Neural Computing*, Van Nostrand, New York.

7.5 MODELING BY FUZZY SETS

Fuzzy Set Theory was introduced by Professor Lotfi Zadeh* of the University of California, Berkeley during the 1960s. Fuzzy set theory accepts many valued logic and departs from the classical logic of Aristotle which allows a proposition to be either true or false. The idea of many-valued logic was developed by Jan Lukasiewicz, a Polish logician in the 1920s and applied by Max Black in 1937. Zadeh formally developed multi-valued set theory in 1965 and called it fuzzy set theory.

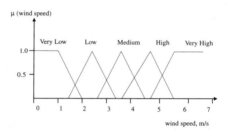

FIGURE 7.25 A fuzzy set of wind speed is represented by four fuzzy levels: "very low," "low," "medium," "high," "very high." By allowing varying degrees of membership, fuzzy sets enables the process of decision making better under uncertainty.

The main idea of fuzzy sets is that they allow partial membership of an element in a set, as shown in Figure 7.25. A fuzzy set **F** in a universe of discourse **U** is defined to be a set of ordered pairs,

$$\mathbf{F} = \{(\mathbf{u}, m_F(\mathbf{u})) | \mathbf{u} \in \mathbf{U}\} \tag{7.19}$$

where $m_F(\mathbf{u})$ is called the membership function of **u** in **U**. When **U** is continuous, **F** can be written as,

$$\mathbf{F} = \int_{\mathbf{u}} \mu_F(\mathbf{u})/\mathbf{u} \tag{7.20}$$

and when **U** is discrete, **F** is represented as,

$$\mathbf{F} = \sum_{i=1}^{n} \mu_F/\mathbf{u} \tag{7.21}$$

where n is the number of elements in the fuzzy set **F**.

---

* Zadeh, L. A., "Fuzzy sets as a basis for a theory of possibility." *Fuzzy Sets and Systems*, 10, (3), 243–260, 1978.

7.5.1 Linguistic Variables

Fuzzy set theory enables modeling a system in the natural language by making use of linguistic variables. A linguistic variable is characterized by a quintuple,

$$(x, T(x), U, G, M)$$

where,

x = name of variable [Example: wind speed]
$T(x)$ = Term set of x [Example: T(wind speed) = {very low, low, medium, high, very high}]
U = Universe of discourse [Example: U(wind speed) = (0, 7) m/s]
G = Syntactic rule for generating the name of values of x
M = Semantic rule for associating a meaning with each value

The terms Very Low, Low, Medium, High, and Very High wind speeds represent fuzzy sets whose membership functions are shown in of Figure 7.25.

If A and B are fuzzy sets with membership function $\mu_A(u)$ and $\mu_B(u)$, respectively, then the membership function of the union, intersection, and complement, and the fuzzy relation involving the two sets, are as follows:

Union (AND)

Example: IF A AND B THEN C

$$\mu_C(u) = \mu_{A \cup B} = \max(\mu_A(u), \mu_B(u)) \quad u \in U$$

Intersection (OR)

Example: IF A OR B THEN C

$$\mu_C(u) = \mu_{A \cap B} = \min(\mu_A(u), \mu_B(u)) \quad u \in U$$

Complement (NOT)

Example: NOT C

$$\mu_{\bar{c}}(u) = 1 - \mu_c(u)$$

Fuzzy Relation

Two or more fuzzy IF/THEN rules of the form,

$$Y \text{ is } B_i \text{ IF } X \text{ is } A_i, \quad i = 1, 2, \ldots n$$

can be connected by the fuzzy relation R,

$$R = \sum_i (A_i \times B_i) \tag{7.22}$$

where the membership function of the cartesian product ($A_i \times B_i$) is given by,

$$\mu_{A_x\ldots A_n}(u_1,\ldots\ldots u_n) = \min\{\mu_{A_1}(u_1),\ldots\ldots,\mu_{A_1}\} \tag{7.23}$$

One of the most successful applications of fuzzy sets is in the design of Fuzzy Logic Controller (FLC). A fuzzy logic controller may be developed to control tranmission line ampacity. The steps in FLC design follow:

- Define input and control variables. For example, in ampacity calculation, the input variables are wind speed and direction, solar radiation, and air temperature. The control variable is ampacity.
- Describe the input and control variables as fuzzy sets (fuzzification).
- Design the rule base (fuzzy control rules).
- Develop the fuzzy computational algorithm and fuzzy output.
- Transform fuzzy outputs to crisp control actions (defuzzification).

An important step in the design of FLC is the selection of membership function and fuzzy IF/THEN rules. Generally, they are obtained by experimentation with data. More recently, neural networks have been used to learn membership function and the rules. Figure 7.26 is a schematic representation of a fuzzy logic controller with learning by neural networks.

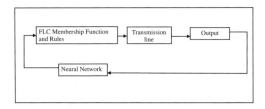

FIGURE 7.26 Schematic representation of fuzzy logic controller that learns from a Neural network.

An example of a fuzzy logic system for the calculation of transmission line ampacity is given in Figure 7.27. Only two rules are shown in the figure for illustration of main concepts, and a calculation to obtain a crisp value of line ampacity from fuzzy sets is given below. This is an excellent example of fuzzy logic because the meteorological variables comprising wind speed and ambient temperature, are best described by fuzzy sets. In the example, wind speed, ambient temperature and ampacity are represented by fuzzy sets having four fuzzy levels (very low, low, medium, high, very high). In this manner, we can represent weather parameters by linguistic variables and consider the uncertainty in weather data comprising wind speed and ambient temperature accurately. Furthermore, weather forecast data from

Weather Modeling for Forecasting Transmission Line Ampacity

the National Weather Service or other sources are normally presented in a similar manner, which can be directly utilized to calculate transmission line ampacity as shown in Figure 7.27.

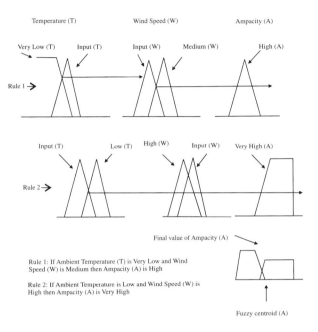

FIGURE 7.27 Fuzzy logic system of calculation of transmission line ampacity.

Example of Ampacity Calculation by Fuzzy Sets

Typical input data for ampacity calculation may be as follows:

Sunny, wind north-west 8–12 km/h, temperature low 2–15°C.

We are required to calculate ampacity from the above data.

The following two rules are activated:

Rule 1: If Ambient Temperature (T) is very low and Wind Speed (W) is medium, then Ampacity (A) is high.
Rule 2: If Ambient Temperature is low and Wind Speed is high, then Ampacity is very high.

The degree of membership of fuzzy variables T and W in the four fuzzy levels is given in Table 1.
From Rule 1, the membership of input T denoted as $m_{VL}(T)$, and the membership of input W denoted as $m_M(W)$, is obtained from Figure 7.27,

TABLE 1

Var/Level	VL	L	M	H	VH
T (2–15) °C	0.75	0.6	0	0	0
W (2.5–3.2) m/s	0	0	0.55	0.75	0

$$m_{VL}(T) = 0.75$$

$$m_M(W) = 0.55$$

Therefore, Rule 1 activates the consequent fuzzy set H of A to degree

$$m_H(A) = \min(0.75, 0.55) = 0.55$$

Similarly, from Rule 2, the membership of input T ($m_L(T)$), and the membership of input W ($m_H(W)$), are obtained from Figure 7.27,

$$m_L(T) = 0.6$$

$$m_M(W) = 0.75$$

Therefore, Rule 2 activates the consequent fuzzy set VH of A to degree

$$m_{VH}(A) = \min(0.6, 0.75) = 0.6$$

Therefore, the combined output fuzzy set of A, $m_o(y_j) = (0,0,0,0.55,0.6)$.

A crisp value of ampacity is obtained by centroid defuzzification of the combined output fuzzy set,

$$A = S_f \frac{\sum_{j=1}^{p} y_j \cdot m_o(y_j)}{\sum_{j=1}^{p} m_o(y_j)}$$

$$y = (1, 2.5, 3.5, 4.5, 5.5)$$

Where the elements y_j of vector y are the mean value of each fuzzy level.

$$A = S_f \frac{0.5 \times 0 + 1.5 \times 0 + 2.5 \times 0 + 4.5 \times 0.55 + 5.5 \times 0.6}{0.55 + 0.6}$$

$$A = S_f \times 5.02$$

Ampacity = 2008 A

Since the membership functions given in Figure 7.26 are based on scaled values the actual value of ampacity for a transmission line having an ACSR Cardinal conductor is obtained by multiplying with a scaling factor, $S_f = 400$.

7.6 SOLAR RADIATION MODEL

During daytime, the transmission line conductor is heated by the energy received from the sun. Depending upon the condition of the sky and the position of the sun with respect to the conductor, the temperature of the conductor may increase by 1–10°C by solar heating alone. To calculate conductor heating by solar radiation, the energy received on the surface of the conductor from the sun (E_s) is obtained as,

$$E_s = \alpha_s(S_b + S_d) \tag{7.24}$$

Where,

α_s = solar absorption coefficient, $0 < \alpha_s < 1$
S_b = Solar energy received by conductor due to beamed radiation
S_d = Solar energy received by conductor due to diffused radiation

The beamed radiation S_b is calculated as,

$$S_b = S_n \cdot \cos(\theta) \tag{7.25}$$

The diffused radiation S_d is calculated as,

$$S_d = S_n \cdot \cos(\theta_z) \tag{7.26}$$

Where,

S_n = the component of solar radiation that is normal to the surface of the earth
θ = Angle of deviation from the normal
θ_z = Zenith angle, given by,

$$\cos(\theta_z) = \sin(\phi)\sin(\delta) + \cos(\phi)\cos(\delta)\cos(\omega) \tag{7.27}$$

The normal component of the beamed solar radiation inside the earth's atmosphere is obtained as,

$$S_{nb} = S_{n(o)} \cdot \tau_b \qquad (7.28)$$

The normal component of the diffused solar radiation inside the earth's atmosphere is obtained as,

$$S_{nd} = S_{n(o)} \cdot \tau_d \qquad (7.29)$$

$S_{n(o)}$ = Normal component of the solar radiation outside the earth's atmosphere which is obtained as,

$$S_{n(o)} = S_c \left\{ 1 + 0.0033 \cdot \cos\left(\frac{360 \cdot J_d}{365}\right) \right\} \qquad (7.30)$$

J_d = 1,2..365 day number
S_c = Solar constant = 1353 W/m² (measured value outside the earth's atmosphere)
τ_b = atmospheric transmittance of beamed radiation. It takes into account attenuation by the earth's atmosphere of the extraterrestrial radiation and is given as,*

$$\tau_b = a_0 + a_1 \exp\left(\frac{-k}{\cos(\theta_z)}\right) \qquad (7.31)$$

τ_d = atmospheric transmittance of diffused radiation given by,

$$\tau_d = 0.2710 - 0.2939\tau_b \qquad (7.32)$$

$a_0 = 0.4237 - 0.00821(6\text{-Alt})^2$
$a_1 = 0.5055 - 0.00595(6.5\text{-Alt})^2$
$k = 0.2711 - 0.01858(2.5\text{-Alt})^2$
Alt = Altitude of the transmission line above MSL, km

The angle of deviation θ of the beamed radiation with respect to the normal to surface of the conductor is given by the following formula,

$$\cos(\theta) = \sin(\delta)\sin(\phi)\cos(\beta) - \sin(\delta)\cos(\phi)\sin(\beta)\cos(\gamma)$$
$$- \sin(\delta)\cos(\phi)\sin(\beta)\cos(\gamma)$$
$$+ \cos(\delta)\cos(\phi)\cos(\beta)\cos(\omega) + \cos(\delta)\sin(\phi)\sin(\beta)\cos(\gamma)\cos(\omega)$$
$$+ \cos(\delta)\sin(\phi)\sin(\beta)\cos(\gamma)\cos(\omega) + \cos(\delta)\sin(\beta)\sin(\gamma)\sin(\omega) \qquad (7.33)$$

---

* Duffie, John A. and Beckman, William A. 1980 *Solar Engineering of Thermal Processes*, John Wiley & Sons, New York.

φ = Latitude. Angle between north or south of the equator, north +ve:

$$-90° =< \phi =< 90°$$

δ = Declination. Angular position of the sun at solar noon with respect to the plane of the equator, north +ve: $-23.5° =< \delta =< 23.5°$. The declination angle is calculated as,

$$\delta = 23.45\left\{360\frac{(284+J_d)}{365}\right\} \quad (7.34)$$

β = Slope. Angle between the conductor axis and the horizontal: $0 =< \beta =< 180°$

γ = Line orientation angle (azimuth). South zero, East -ve, West +ve: $-180° =< \gamma =< 180°$

ω = Hour angle. Angular displacement of the sun east or west of the local meridian due to rotation of the earth on its axis at the rate of 15°/hr, morning -ve, afternoon +ve. The hour angle is given by,

$$\omega = (12 - \text{SolarTime}) \cdot 15 \quad (7.35)$$

$$\text{SolarTime} = \text{StandardTime} + 4(L_{std} - L_{loc}) + \text{Eqt} \quad (7.36)$$

L_{std} = Longitude of Standard Meridian (Example: L_{std} San Francisco = 120°)
L_{loc} = Longitude of location

$$\text{Eqt} = \text{Equation of time} = 9.87\sin(2B) - 7.53\cos(B) - 1.5\sin(B) \quad (7.37)$$

$$B = \frac{360(J_d - 81)}{364} \quad (7.38)$$

θ = Angle of incidence. The angle between the beam radiation on a surface and the normal to the surface. These angles are shown in Figures 7.28 and 7.29. The result of solar radiation calculation by the program is for one day during the month of July in the region of San Francisco and is shown in Figure 7.30.

Figure 7.31 is a flow chart showing the line ampacity forecasting procedure from weather models AmbientGen, WindGen, and SolarGen.

7.7 CHAPTER SUMMARY

In this chapter, weather modeling for the prediction of transmission line ampacity is presented firstly by Fourier analysis of weather data. Ambient temperature and

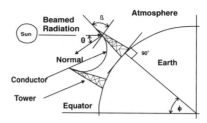

FIGURE 7.28 Transmission line solar angles.

FIGURE 7.29 Slope angle between conductor and horizontal.

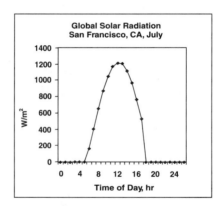

FIGURE 7.30 Global Solar Radiation (Direct Beam + Diffused Radiation) on a transmission line conductor surface in San Francisco, calculated by program for one day during the month of July. Transmission Line E-W direction and the slope angle is 5 °.

wind speed models were developed by fitting Fourier series to hourly weather data available from the National Weather Service. A Kalman filter-type algorithm is then used to model the uncertainty in the Fourier series, and a real-time forecasting algorithm is presented that uses a recursive estimation procedure for the prediction of ambient temperature and wind speed. The forecasts are adapted to the daily high and low values of ambient temperature that are forecast daily by the National Weather Service.

Weather Modeling for Forecasting Transmission Line Ampacity 141

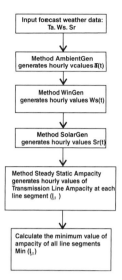

FIGURE 7.31 Flow chart for forecasting transmission line ampacity from weather models.

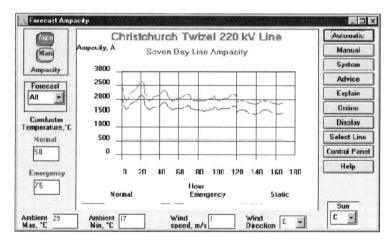

FIGURE 7.32 Forecasting hourly ampacity values of a 220 kV transmission line seven days in advance in the region of North Island, New Zealand.

Neural network models for prediction and weather pattern recognition are developed by using the back propagation algorithm, and a self-organizing neural network is developed using Kohonen's learning algorithm. Neural networks offer an alternative method of forecasting weather variables and pattern recognition.

Basic fuzzy logic concepts are presented and a system for developing a fuzzy logic controller of transmission line ampacity is proposed for further research. Analytical expressions for the calculation of hourly values of solar radiation are also developed which take into account transmission line location and line geometry.

For real-time prediction of transmission line ampacity, a recursive estimation algorithm for weather forecasting is developed based on Kalman filter equations. Examples of weather models developed by the program are shown for the region of San Francisco.

8 Computer Modeling

8.1 INTRODUCTION

For the proper evaluation of transmission line capacity, it is necessary to support the theory developed in Chapter 4 by a practical knowledge of the transmission system and its environment. Line ampacity is calculated from steady, dynamic, and transient thermal models by developing an object model and expert rules of the transmission line system. This is the object of computer modeling of line ampacity system as described in this chapter.

In this chapter the LINEAMPS (Line Ampacity System) transmission line expert system computer program is developed by the integration of theory and practical knowledge of the transmission line system. Examples of object-oriented modeling and expert system rules are presented here to demonstrate how practical knowledge is embedded in program, which will enable a transmission line engineer to easily evaluate power line capacity in any geographic region.

8.1.1 FROM THEORY TO PRACTICE

A transmission line is composed of the conductors that carry current, structures that support conductors in air, insulators to safely protect transmission tower structures from the high voltage conductor, connectors for the splicing of conductors, and other hardware necessary for the attachment of conductors to the tower. The transmission line environment is vast as they traverse different kinds of terrain—plains, forests, mountains, deserts, and water. They are also exposed to varying atmospheric conditions—sun, wind, temperature, and rain. Some sections of a line may be exposed to industrial pollution as well. All of these environmental factors affect line capacity to a certain degree. Modeling such a system is not easy. A systematic approach using practical knowledge and simplifying assumptions is required to achieve a realistic line ampacity system with sufficient accuracy. This is the object of the line ampacity expert system program.

8.1.2 THE LINEAMPS EXPERT SYSTEM

LINEAMPS* is an expert system computer program developed by the author based on a systematic approach of object-oriented modeling and expert rules. Objects are used to model the transmission line system and its environment. Expert system rules are used to incorporate practical knowledge. The end product is an intelligent line ampacity system resembling a human expert. It is also a humble contribution and a

---

* Anjan K. Deb, Object oriented expert system estimates transmission line ampacity, *IEEE Computer Application in Power*, Volume 8, Number 3, July 1995.

practical demonstration of research in the field of artificial intelligence,* where a computer system demonstrates intelligent behavior. In the following sections the object-model and expert system design of the line ampacity system are presented in greater detail.

8.2 OBJECT MODEL OF TRANSMISSION LINE AMPACITY SYSTEM

System modeling by object orientation is a new way of data representation and programming.** Its attributes and behavior define an object. Objects can store data, whether it is temporary or permanent. Methods or stored procedures in the objects give them behavior, which enables them to perform a required action. Methods have the ability to use data contained in their own objects as well as other objects. For example, data stored in weather objects of the line ampacity system program are environmental data comprising weather conditions, terrain, latitudes, longitude, and elevation. Data relating to the electrical and mechanical properties of the transmission line are contained in the transmission line object.

Once an object is created it is easier to create newer instances of the same object by inheritance. Objects inherit data as well as methods—this is an important property of all objects. For example, once a transmission line object is created, several lines may be produced by inheritance. Similarly, weather station objects are created. These objects have methods to receive weather forecast data from external sources, for example, the National Weather Service, the Internet, or even the daily newspaper. By using forecast weather data, as well as the stored procedures and weather patterns of the region, hourly values of transmission line ampacity are forecast several hours in advance. The object model of the line ampacity system is given in Figure 8.1, and the hierarchical structure of transmission line, weather station, and conductor object of the transmission line ampacity system is shown in the Figures 8.2, 8.3, and 8.7.

8.2.1 LINEAMPS OBJECT MODEL

The object model of the LINEAMPS system shown in Figure 8.1 is comprised of transmission line object, weather station object, conductor object, and cartograph object. In addition, there is a system of objects for the presentation of data, and a user-friendly graphical user interface. The Kappa-PC*** object oriented modeling tool is used to implement the object model. The following sections describe the object model and expert rules of the line ampacity system.

---

* Lawrence Stevens, Artificial Intelligence. The Search for the Perfect Machine, Hayden Book Company, 1993.
** G. Booch, Object-Oriented Design with Applications, Benjamin Cummings Publishing Co. 1991.
*** Kappa-PC® object-oriented development software v 2.4, 1997, Intellicorp, Mountainview, CA.

Computer Modeling

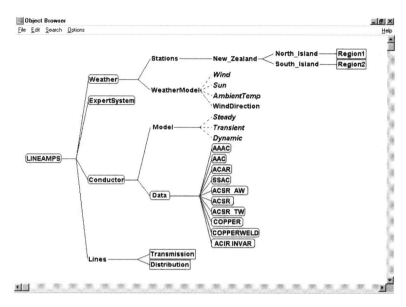

FIGURE 8.1 Line ampacity system object-model.

8.2.2 Transmission Line Object

The Line object class is shown in the Figure 8.2. Powerline objects are classified by categories of transmission and distribution lines. In each class there are subclasses of lines by voltage levels. In each subclass there are several instances of powerlines, for example, the transmission line object class is comprised of 500kV, 345kV, and 230kV lines. The distribution line class comprises 15kV, 480 V, and other line instances created by the user. Each of these transmission line objects derives its attributes and behavior from the general class of lines.

The subclasses of line voltages in the line object classes are defined by the user. The line object has all the data and methods pertaining to the overhead line that are necessary for the evaluation of powerline ampacity. Data are stored in slots, and methods perform the action of evaluating ampacity. Table 8.1 shows an example of the data in one instance of a transmission line object. Only a partial list of attributes and methods are shown for the purpose of illustration.
Following is a complete list of attributes of the line object:

Line object attributes:

LineName
LineVoltage
LineLength
ConductorCodeName
ConductorType
ConductorDiameter

ConductorArea
ConductorAlpha
ResistanceAtTemperature
Conductor DC Resistance
ConductorEmissivity
ConductorAbsorbtivity
ConductorSpecificHeat
ConductorMass
ConductorAluminumMass
ConductorSteelMass
NumberOfSites
SiteList (List)
AssociatedWeatherSites(List)
Site#x(List), where x =1,2..NumberOfSites
AmbientTemperatureSite#x(List)
WindSpeedSite#x(List)
WindDirectionSite#x(List)
SkyConditionSite#x(List)
NormalAmpacityOneDay(List)
NormalAmpacitySevenDays(List)
EmergencyAmpacityOneDay(List)
EmergencyAmpacitySevenDays(List)
TimeOfDayEnergyPrice(List)

Each Site#x comprises a list having the following values: elevation, slope, latitude, longitude, standard longitude, line orientation at the location, and the type of terrain.

Line object comprise the methods shown in Table 8.2.

TABLE 8.1
Transmission Line Object: 350kV_Line10

Attributes	Values	Method
Name	San Francisco, Berkeley	Ampacity
Distance	50 km	Steady_State_Ampacity
Conductor	acsr cardinal	Dynamic_Ampacity
Duration	30 min	Transient_Ampacity
Region	Coastal	Draw Line

A graphical representation of transmission line by voltage category and by the type of line, transmission or distribution, greatly facilitates the task of a transmission line engineer to view and modify line data by simply clicking on a transmission line object shown in Figure 8.2.

The following is an example of creating a transmission line object, assigning attributes and giving them behavior.

Computer Modeling

TABLE 8.2

Method:	Function:
Amp7Days	Calculates hourly values of line ampacity up to seven days in advance.
SteadyStateCurrent	Calculates steady state current.
SteadyStateTemperature	Calculates steady state temperature.
DynamicAmpacity	Calculates conductor temperature response due to step change in line current.
TransientAmpacity	Calculates conductor temperature response due to short circuit and lightning current.
DrawLine	Draws the line in the cartogram window.
CheckInput	Verifies the correctness of input data.
WeatherData	Obtains data from the associated weather stations.
AdjustWeather	Weather data is adjusted in AmbientTemperatureSite#x slot and WindSpeedSite#x slot based on terrain in Site#x slot.
MakeNewLine	Makes an instance of a new line.
UpdateLineList	Updates the list of lines when a new line is created.
MakeSiteData	Makes virtual weather sites along the route of the line.
EnergyDeliveryCost	Calculates hourly values of energy delivery cost based on time of day energy price and forecast ampacity.

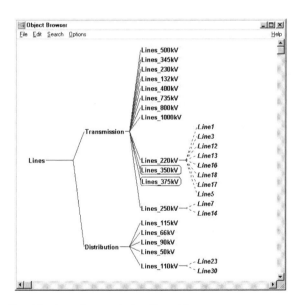

FIGURE 8.2 Classification of transmission line objects.

Creating the object

    MakeClass(Transmission, Lines);
    MakeInstance(350kV_Line10, Transmission);

Attributes

    MakeSlot(350kV_Line10, Name, SanFrancisco_Danville, 50);
    MakeSlot(350kV_Line10, Distance, 50);
    MakeSlot(350kV_Line10, Conductor, Cardinal);

Behavior

    MakeMethod(Transmission, Ampacity, [time, ambient, sun, wind, interval]);

Action

    SendMessage(350kV_Line10, Ampacity, [12:00, 20, C, 2, 60]);

Result:

    1000 A

8.2.3 WEATHER STATION OBJECT

Since LINEAMPS calculates transmission ampacity from weather data, modeling of weather by developing an object model is an important aspect of this program. The purpose of the weather object is to reproduce, as closely as possible by software, the behavior of an actual weather station. This is the main objective of the LINEAMPS weather station object. The behavior of a weather station object is obtained by modeling weather patterns of the region by Fourier analysis from historical weather data and regional weather forecasts prepared daily by the National Weather Service.

To enable modeling of a weather station object, it is divided into subclasses of regions and region types so that weather station instances inherit class attributes. Station objects have all of the meteorological data and geographic information required in the calculation of transmission line ampacity. The weather station object hierarchy is shown in Figure 8.3 and is comprised of:

1. Subclass of regions. Example: Region1, Region2... Region#x.
2. Subclass of region types. Example: Coastal, Interior, Mountain, Desert.
3. Instances of weather stations. Example: San Francisco, Oakland, Livermore.

Weather station objects have the following attributes.

Attributes of weather station object

- StationName
- AmbientMax(List)

Computer Modeling

- AmbientMin(List)
- HourAmbientMax
- HourAmbientMin
- AmbientPattern#x(List), where x =1,2...12 months
- WindSpeedMax(List)
- WindSpeedMin(List)
- HourWindSpeedMax
- HourWindSpeedMin
- WindSpeedPattern#x(List), where x =1,2...12 months
- WindDirection(List)
- SkyCondition(List)
- ForecastTemperature(List)
- ForecastWind(List)
- ForecastSolarRadiation(List)
- Latitude
- Longitude

A "List" inside parentheses is used to indicate that the attribute has a list of values. Weather station objects comprise the following methods:

TABLE 8.3

Method	Function
AmbientGen	Generates hourly values of ambient temperature, (Figure 8.4).
WindGen	Generates hourly values of wind speed, Figure 8.5.
SolarGen	Generates hourly values of solar radiation, Figure 7.3
SelectPattern	Selects ambient temperature and wind speed pattern of the month
DisplayAmbient	Display ambient temperature in a line plot and a transcript image
DisplayWind	Display wind speed in a line plot and a transcript image
OnLineData	Reads weather data downloaded from America-On-Line.
MakeNewStation	Makes an instance of a new weather station.

NEW ZEALAND EXAMPLE*

To fix ideas, an example of New Zealand weather station object is presented in Figure 8.3. The object has North and South Island subclasses. North Island is Region 1, and South Island Region 2. Each region is further divided into subclasses of Coastal, Interior, Mountain, and Desert. In each subclass there are instances of weather stations. These instances derive their attributes and behaviors from the general class of weather stations, and their characteristics are refined by the properties of each region

Table 8.4 shows data in one instance of a weather station object.

---

* Deb, Anjan K., *LINEAMPS for New Zealand, A Software User's Guide*, 1996.

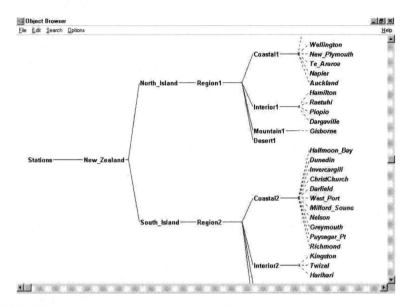

FIGURE 8.3 Classification of weather station objects.

TABLE 8.4
Weather Station Object: Wellington Example

Attribute	Value	Method
Name	Wellington	Ambient_Gen
Latitude	41° 18 ' S	Wind_Gen
Longitude	174° 47' E	Solar_Gen
Elevation	100	Draw_on_Map
Region	Coastal	Show_Value

Examples of Ambient_Gen and Wind_Gen methods used to generate hourly values of ambient temperature and wind speed are shown in Figures 8.4 and 8.5. Seven days' forecast weather data from the National Weather Service are shown in Figure 8.6.

8.2.4 Conductor Object

A conductor object class is shown in Figure 8.7. It is comprised of the following sub-classes of conductor types: AAAC, AAC, ACAR, SSAC, ACSR_AW, ACSR, ACSR_TW, COPPER, COPPERWELD and ACSR_INVAR. Each conductor type has plurality of conductor instances. The user may also create other subclasses of conductor types and new instances of conductor objects.

Computer Modeling 151

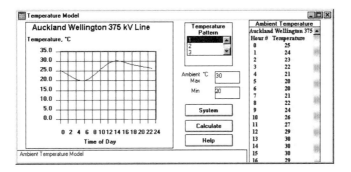

FIGURE 8.4 Temperature modeling.

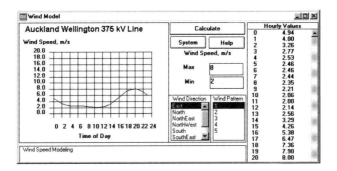

FIGURE 8.5 Wind speed modeling.

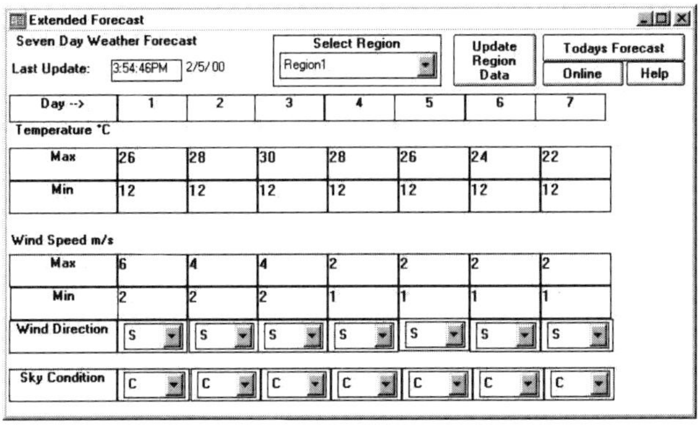

FIGURE 8.6 National Weather Service seven day weather forecast.

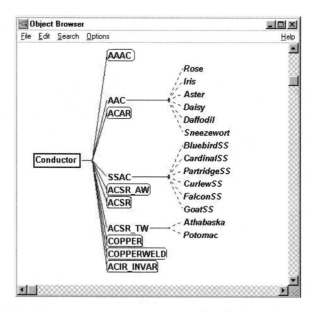

FIGURE 8.7 Classification of transmission line conductor objects.

Conductor object has following attributes:

Attributes of Conductor object:

Conductor code name
Conductor type
Conductor diameter
Conductor area
DC resistance of conductor
Emissivity of conductor
Absorptivity of conductor
Specific heat of conductor
Conductor mass
Aluminum mass
Steel mass

Conductor object has following methods:

Data pertaining to one instance of a transmission line conductor is shown in Figure 8.8.

8.2.5 Cartograph Object

A cartograph window is used to show the location of weather stations and the transmission line route in a geographic map of the region, as seen in Figure 8.9.

Computer Modeling

TABLE 8.5

Methods	Function
SpecificHeat	Calculates the specific heat of conductor
SteadyStateCurrent	Calculates steady state current
SteadyStateTemperature	Calculates steady state temperature
DynamicAmpacity	Calculates conductor ampacity n the dynamic state.
DynamicTemperature	Calculates conductor temperature versus time in the dynamic state.
TransientAmpacity	Calculates conductor temperature versus time in the transient state.
MakeNewConductor	Makes an instance of a new conductor

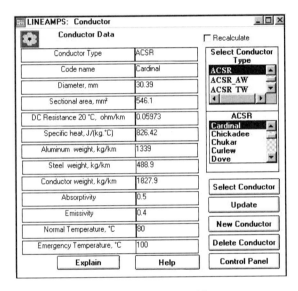

FIGURE 8.8 Data in a transmission line conductor object.

The maximum and minimum ambient temperatures of the day are also displayed at the location of each weather station. By displaying the transmission line in a map, one obtains a better picture of the transmission line route and its environment.

A unique feature of the program realized by object-oriented modeling is the ability to create new lines and weather stations by inheritance. A transmission line object is created by entering the latitudes and longitudes of the line at discrete intervals, and by specifying the type of terrain through which the line passes. Similarly, weather station objects are also created. A transmission line appears on the map when the line is selected from the database. It is generated automatically by the program with a DrawLine method using data stored in the transmission line object shown in Figure 8.9.

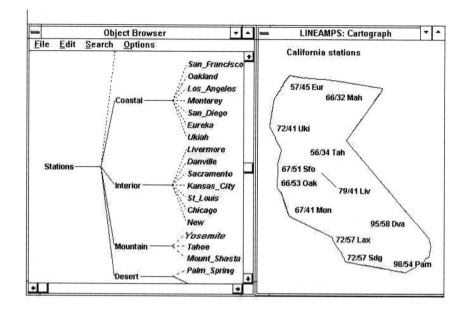

FIGURE 8.9 A cartograph is shown in the right window of the LINEAMPS program and the corresponding weather station objects are shown in the left window. The cartograph shows the geographic boundary of the region that was created by program. The location of weather stations, daily maximum and minimum values of ambient temperature (max/min) and the trace of a selected 230 kV transmission line from Sacramento to Livermore, California USA is shown in the cartograph window.

8.3 EXPERT SYSTEM DESIGN

The line ampacity expert system is accomplished by a system of rules and goals to be achieved by the program. Expert systems are capable of finding solutions to a problem by a description of the problem only. The rules and the data in the objects describe the problem. This declarative style of rule based programming offers an alternative to the traditional procedural programming method of solving problems. For example, to solve a problem by rules, we specify what rules to apply and a goal. Reasoning is then carried out automatically by an inference engine, which finds a solution by using a backward or forward chaining mechanism.*

An expert system is generally composed of the following:

- Goals, facts, database
- Rules or knowledge base
- Inference engine (reasoning capability)
- Explanation facility
- Man machine interface
- Learning capability

---

* Waterman, Donald A., *A Guide to Expert Systems*. Addison-Wesley, Reading, MA, 1986.

Computer Modeling

In the following section, the transmission line expert system is described by presenting an example of goal-oriented programming, rules, inference engine, and explanation facility. These features were used in the program to check user input data and explain error messages. An example of a man machine interface was given in a previous IEEE publication.* Learning by artificial neural network is described later in this chapter.

8.3.1 GOAL-ORIENTED PROGRAMMING

Goal-oriented programming by rules greatly facilitates the task of computer programming as the programmer is not required to code a detailed logic to solve a problem. In traditional programming by procedures, a programmer must precisely code the logic, of a mathematical equation for example, to solve a problem. Rules not only facilitate a declarative style of programming, but also provide a practical method of incorporating practical and imprecise knowledge such as "rules of thumb" that are not easily amenable to formal mathematical treatment. In addition, rules are easily understood and maintained. Following is a simple example of programming by rules and a goal in the line ampacity system.

Goal: SteadyStateGoal

Action:

```
{ SetExplainMode( ON );
ForwardChain( [ ASSERT ], SteadyStateGoal, Global:SteadyStateRules );
If ( Transmission:Problem # = N )
Then CalcSteadyTemperature( );
};
```

The object of the above action statement is to satisfy a Steady-State Goal by verifying all of the steady-state rules stored in the Object:Slot pair Global:SteadyStateRules. In the Kappa-PC object-oriented development environment, slots are provided to store data of an object. The program proceeds with the calculation of transmission line conductor temperature only if there are no problems detected in the data entered by the user. Setting the explain mode to ON enables the user to receive explanations of expert system generated error messages. When new facts are generated by the firing of rules, [ASSERT] ensures that the new facts are automatically inserted into a fact database.

Result:

In the above example, the user input data were checked by the expert system rules. The SteadyStateGoal was satisfied and the program correctly evaluated steady-state conductor temperature to be equal to 60°C.

In the following example, the user entered a value of conductor temperature less than ambient temperature. The expert system correctly detected the problem and

---

* Deb, Anjan, K., Object oriented expert system estimates transmission line ampacity, *IEEE Computer Application in Power*, Volume 8, Number 3, July 1995.

generated an error message, as shown in Figure 8.10. By clicking on the Explain button, the expert system generated the required explanations, and the story of the transmission line ampacity problem started to unfold (Figure 8.11).*

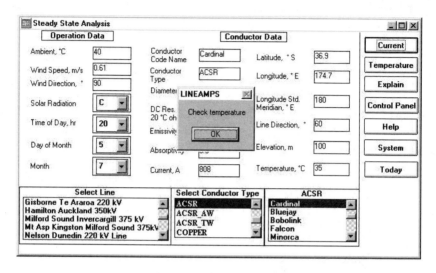

FIGURE 8.10 Example of error message given by program when user entered incorrect value of conductor temperature.

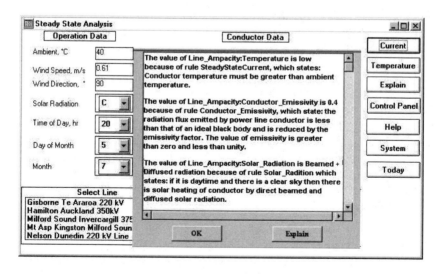

FIGURE 8.11 Explanation of error message given by program when user clicks on the explain button.

---

* *Towards the Learning Machine*, Richard Forsyth.

Computer Modeling 157

8.3.2 Expert System Rules

The expert system knowledge base is comprised of the abovementioned system of objects and rules. In LINEAMPS, rules are used to offer expert advice to users in the event of erroneous input or conflicting data, or to caution the user during specific operating conditions. Some examples of rules used in the program are:

Rule 1.
If ambient temperature is greater than conductor temperature, then advise user.

Rule 2.
If the temperature of the selected conductor is greater than the allowable maximum for the conductor type, then advise user.

Rule 3.
If user input, is low 2 or 4 ft/s wind speed, and the National Weather Service forecast is high wind speed, then advise user.

Rule 4.
In the dynamic state, if preload current results in a higher than maximum allowable conductor temperature, then advise user.

Rule 5.
In the dynamic state, if post overload current results in a higher than maximum allowable conductor temperature, then advise user.

Rule 6.
In the dynamic state, if the user specified preload current results in a conductor temperature that is higher than the allowable maximum, then advise user.

Rule 7.
In the transient state, if the duration of transient current is greater than the specified maximum, then advise user.

Rule 8.
If the age of the conductor is old and the conductor temperature is high, then advise user.

Rule 9.
If the age of the conductor is old or the line passes through areas of industrial pollution, and the coefficient of solar absorbtivity is low and/or the emissivity of the conductor is high, then advise user.

Rule 10.
In the transient state, if the transient current is high and the line is old, then advise user.

Rule 11.
If the line passes through urban areas with high-rise buildings or where wind is restricted by tall structures or trees and line ampacity is high, then advise user.

Rule 12.
If the value of conductor emissivity is less than or equal to 0 or greater than 1, then advise user.

Rule 13.
If the value of conductor absorbtivity is less than or equal to 0 or greater than 1 then advise user.

The following example shows a listing of two rules used by the program.

```
/******************************************************
**** RULE 1: Conductor Rule
**** If conductor temperature is greater than the allowed
**** maximum for the conductor type, then reduce current
******************************************************/
MakeRule( ConductorRule, [],
Not( Member?( ConductorTypes:AllowableValues, Steady:ConductorType ) )
Or Steady:Temperature > Steady:ConductorType:MaxTemperature,
{
Transmission_Line_Ampacity:Problem = "High conductor temperature";
PostMessage( "Please check valid conductor type and conductor temper-
ature" );
} );
SetRuleComment( ConductorRule, "If conductor temperature is greater
than the
allowed maximum for the conductor type, then reduce current" );

/******************************************************
**** RULE 2: Steady-State Temperature
**** Conductor temperature must be greater than
**** ambient temperature.
******************************************************/
MakeRule( SteadyStateTemperature, [],
Steady:Calculate #= Current And
Steady:ShowTemperature <= Steady:AmbientTemperature,
{
Line_Ampacity:Problem = "Conductor temperature";
PostMessage( "Check ambient temperature" );
} );
SetRuleComment( SteadyStateTemperature,
"Conductor temperature must be greater than ambient temperature");
```

In the above steady-state analysis window of the LINEAMPS program, the value of conductor temperature entered by the user is 35°C, which is inconsistent with the value of the ambient temperature, 40°C. One of the rules in the expert system detects this problem and displays the error message shown in Figure 8.10.

After receiving an explanation, users also have the ability to request other relevant information regarding various input data required in this session window. In Figure 8.11 explanations of temperature, emissivity, and solar radiation are given.

8.4 PROGRAM DESCRIPTION

8.4.1 LINEAMPS Windows

In this section, the main features and functions of the LINEAMPS program are briefly described. LINEAMPS is designed as a system of windows where users conduct various sessions with the program on different aspects of power line ampacity. Following are the main session windows that are presently available in the program for the analysis and planning of transmission line ampacity.

- LINEAMPS Control Panel
- Steady-State Analysis
- Dynamic Analysis
- Transient Analysis
- Forecast Ampacity
- Power Lines
- Conductors
- Weather

8.4.2 Modeling Transmission Line and Environment

The following additional session windows are available for the modeling of a transmission line and its environment, conductor modeling, weather modeling, and cartograph. These windows may be opened by clicking icons, or by selection from the LINEAMPS Control Panel Window.

- Power Lines
- Conductor
- Static Rating
- Cartograph
- Ambient Temperature
- Wind Speed
- Daily Weather Forecast
- Extended Weather Forecast
- Object Browser

The functions and operations of the LINEAMPS program in each window are described in detail in the user manual.*

8.4.3 LINEAMPS Control Panel

The program is operated by a system of icons, each representing a unique function. While full details of the program is presented in the user manual, a brief description of each icon shown in the Control Panel window (Figure 8.12) is presented here.

The control panel window has the following icons:

---

* *LINEAMPS User Manual*, 1998.

FIGURE 8.12 LINEAMPS Control Panel window.

Steady-State Analysis

Clicking this button opens the Steady-State Analysis window. It is used for the analysis and display of transmission line ampacity and conductor temperature in the steady state.

Dynamic Analysis

Clicking this button opens the Dynamic Analysis window. It is used for the analysis and display of transmission line ampacity and conductor temperature in the dynamic state.

Transient Analysis

Clicking this button opens the Transient Analysis window. It is used for the analysis and display of transmission line ampacity and conductor temperature in the transient state.

Forecast Ampacity

Clicking this button opens the Forecast Ampacity window. It is used for the calculation and display of hourly values of transmission line ampacity up to seven days in advance.

Extended Weather Forecast

The Extended Weather Forecast window is opened by clicking this button. It is used to view the forecast weather data of the region, as well as for input of forecast weather data.

Ambient Temperature Model

Clicking this button opens the Ambient Temperature Model window. It is used for selecting weather patterns and for generating hourly values of ambient temperature data from forecast weather data.

Wind Speed Model

Clicking this button opens the Wind Speed Model window. It is used for selecting weather patterns and for generating hourly values of wind speed data from forecast weather data.

Cartograph

Clicking this button opens the Cartograph window. It is used to view a geographic map of the region, the location of weather stations, and the transmission line route.

Powerlines

Clicking this button opens the Powerlines window. This window is used for creating new lines, and viewing and updating data in existing line objects.

Conductor

Clicking this button opens the Conductor window. It is used for conductor selection, creating new transmission line conductors, and viewing data in conductor objects.

Transmission Cost

Clicking this button opens the hourly Transmission Cost window. This window is used to generate hourly values of transmission costs for seven days in advance.

Welcome

The Welcome window is opened by clicking this button. It is used for navigating the LINEAMPS system of windows.

Help

Clicking this button opens the Help window. This window is used for generating help text on user-requested topics.

Print

Clicking this button opens the Print window. It is used to print LINEAMPS-generated reports on steady-state analysis, dynamic analysis, and transient analysis, and forecast hourly values of ampacity for seven days in advance.

Exit

Exit the program by clicking this button.

8.5 CHAPTER SUMMARY

This chapter describes computer modeling of the transmission line ampacity system by objects and rules. The theory developed in the previous chapter and practical knowledge of the transmission line system are implemented in the LINEAMPS program by object-oriented modeling and expert rules.

The object model of the complete line ampacity system is presented, followed by the component object models consisting of transmission line object, weather station object, conductor object, and a cartograph object. The creation of objects, and methods and the sending of messages between objects are presented to show how an elegant system of objects having messaging capability is realized in the LINEAMPS program.

A combination of procedures, goals, and rules are used to find a solution to the powerline ampacity problem. Procedures or methods are used in objects when a mathematical model is available. Rules are used to incorporate practical knowledge. Decisions based on rules are generated automatically by an inference engine. Examples are presented to show how the system calculates powerline ampacity by objects and rules, checks user input, and explains error messages to the user, thus demonstrating the intelligent behavior like a true expert.

A brief description of the various functions and features of the program and the graphical user interface are presented to demonstrate the realization of a complete line ampacity system that is suitable for all geographic regions.

9 New Methods of Increasing Transmission Capacity

9.1 INTRODUCTION

AC transmission circuits are mostly composed of passive elements having very little controllability. Therefore, when the load increases, existing network control methods are not sufficient to properly accommodate increased power flows. As a result, certain lines are more heavily loaded and stability margins are reduced. Due to the difficulty of controlling power flows by existing methods, new types of power electronic devices called FACTS (Flexible AC Transmission System) are used to control existing T&D networks. FACTS devices are installed at all voltage levels up to 800 kV.

FACTS devices are also used in low-voltage distribution networks. Due to greater utilization of electrotechnologies in the industry, and increasing use of high technology electronic equipment like computers, TV, and other similar devices in homes and businesses, consumers are paying greater attention to power quality in distribution circuits. FACTS devices are, therefore, used in power distribution networks to maintain power quality, and voltage regulation, and to lower harmonics and minimize voltage flicker.

This chapter presents an overview of FACTS and the various power semiconductor devices it uses. When transmission capacity increases are planned by dynamic thermal ratings and reconductoring of existing circuits with special or higher-size conductors, it will be necessary to properly evaluate their impact on power system performance and load flows. Installation of FACTS devices may be required for better control of the current and voltage in a transmission line. It is hoped that by gaining knowledge of the various FACTS technologies presented in this chapter, a power transmission and distribution line engineer will make better decisions regarding the selection of FACTS technology best suited for his or her requirements.

9.2 ADVANCEMENT IN POWER SEMICONDUCTOR DEVICES

Recent advances in power electronics semiconductor devices have made it possible to develop equipment with sufficient current and voltage ratings to enable their utilization in electric power circuits for better control of voltage and current. An important application of power electronic semiconductor devices in electric networks

FIGURE 9.1 Thyristor

is to control the amount of current that can pass through the device in response to a control action. The most common semiconductor device used for the control of current flowing through a circuit is a silicon-controlled rectifier device called a "thyristor," shown in Figure 9.1.

THYRISTOR

A thyristor is a three-terminal device with an anode, cathode, and gate, as shown in Figures 9.2-9.5. It is a special type of diode. Like a diode, a thyristor requires a certain positive anode-to-cathode voltage, but unlike a diode it also requires a pulse having a certain voltage and current to be applied to the gate to turn on the thyristor for conduction to begin from anode to cathode. Similar to a diode, conduction takes place during the positive half of the AC cycle, when the anode-to-cathode voltage is positive. It is blocking during the negative half of the AC cycle when the anode-to-cathode voltage becomes negative.

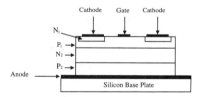

FIGURE 9.2 Thyristor Construction

The current through a thyristor is regulated by controlling the gate-triggering time. This is accomplished by supplying a current pulse to the gate at a desired triggering time. The ability to control current in a circuit by controlling the firing angle, θ, of a thyristor is illustrated by a simple circuit in Figure 9.6.
Considering a resistive load, R, the instantaneous load current, i_r, is

New Methods of Increasing Transmission Capacity

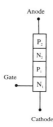

FIGURE 9.3 Schematic

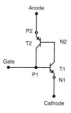

FIGURE 9.4 Two transistor model of Thyristor

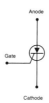

FIGURE 9.5 Thyristor Symbol

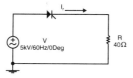

FIGURE 9.6 Current control by Thristor in a 5 kV line to ground distribution circuit.

$$i_r = \frac{V \sin(\omega t)}{R} \quad 0 \leq \omega t \leq \pi$$

The average half-wave rectified DC load current, I_r, is

$$I_r = \frac{1}{2\pi} \int_\theta^\pi \frac{V \cdot \mathrm{Sin}(\omega t)}{R} d\omega t$$

$$I_r = \frac{V \cdot (1 + \mathrm{Cos}(\theta))}{2\pi R}$$

where θ is the angle at which a pulse is applied to the thyristor gate.

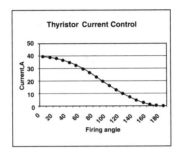

FIGURE 9.7 Control of circuit current as a function of thyristor firing angle.

The variation of load current as a function of the firing angle, θ, in the circuit of Figure 9.6 is shown in Figure 9.7.

The main disadvantage of a thyristor is that it cannot be turned off easily by applying a control signal at the gate. This limitation is overcome by the newer type of semiconductor switching devices like the IGBT discussed later. Other than the thyristor, there are several types of semiconductor devices that are suitable for various power system applications. These are primarily electronic switching devices derived from transistor technology and adapted for high-current and high-voltage applications. Their important characteristics are listed in the Table 9.1 and are briefly discussed here.

MOSFET

The MOSFET is a voltage-driven source used mainly in low-voltage applications. It can be turned on or off rapidly and is suitable for high switching-frequency operations. Being a low voltage device, it is not suitable for direct connection to a high-voltage network, and is therefore used in low-voltage FACTS control circuits, for example, to control a GTO device.

GTO

The Gate Turn Off Thyristor (GTO) is turned on by a short pulse of gate current, and is turned off by applying a reverse gate signal. It has a short turn-off time in the order of tens of nanoseconds, which is much faster than a thyristor. The main disadvantage of a GTO is the high current requirement to turn off current. It has a low ratio of commutation current to turn off current, generally in the range of 3 to 5. This reduces the power that can be commuted by a GTO in a FACTS application.

IGBT

The IGBT takes advantage of the high commutation speed of a power MOSFET and the low resistance offered by a bipolar transistor. The term "insulated gate" refers to the metal oxide insulated gate which requires very little control power. This feature is particularly useful in high-voltage FACTS applications because the control signal is transmitted from ground potential to the gate of the IGBT device at very high voltage. The switch can be turned on or off quite rapidly by the application of a control signal at the gate. It is therefore suitable for high-voltage applications, where several of these devices can be connected in series.

GBTR

The Giant Bipolar Transistor (GBTR) is older than the IGBT, but is available with sufficient power capability that is comparable to a thyristor. Even though the cost of the GBTR is somewhat less than IGBT and power MOSFET, it is not widely used due to the complex electronics required to control it.

MCT

The MOS-Controlled Thyristor is considered the power semiconductor switching device of the future. The MCT device under development is expected to offer high commutation power similar to a GTO, with control capability similar to an IGBT.

FACTS Semiconductor Valve Assembly

Since the present state of the art in semiconductor assembly allows a thyristor or an IGBT device to be built with voltage rating up to 10 kV, several of these devices are required to be connected in series to withstand high commutation voltage for high-voltage applications. Therefore a thyristor valve assembly is made modular in structure for ease of installation and maintenance. At the individual thyristor level it is comprised of a control unit for controlling the firing angle of the thyristor, a unit for cooling the thyristor, and an electrical filter for the elimination of noise generated by switching action. A module may consist of four to six thyristors connected in series with a reactor. Several of these modules are then connected in series to develop the full transmission-line-to-ground voltage.

A schematic of a thyristor valve assembly is shown in Figure 9.8. A thyristor valve assembly is comprised of a system for the communication and control of the various thyristor units, and a system for the distribution of cooling fluid to the various units and support insulators. High-voltage DC convertors and SVC stations are generally located indoors, where a stack of thyristor modules forming a high-voltage valve is either floor standing or suspended from a ceiling.
A number of such devices have been developed which are widely known as FACTS (Flexible AC Transmission System) devices. By connecting these active devices at

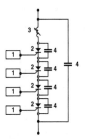

FIGURE 9.8 Construction of a thyristor valve module.

TABLE 9.1
Semiconductor Device Properties

Semiconductor Device	Rated Voltage	Rated Current	Speed	Switching Frequency
MOSFET	200 V	100 A	200 ns	100 kHz
IGBT	1200 V	300 A	1 μs	10 kHz
GBTR	1200 V	300 A	5 μs	3 kHz
GTO	4000 V	3000 A	40 μs	1 kHz
Thyristor	5000 V	4000 A	–	300 Hz
MCT	3000 V	30 A	40 μs	1 kHz

Adapted from "Technologie des FACTS," Ph. Lataire, SEE Conference 1994.

suitable locations in the T&D network, it is now possible to implement a wide range of control over an AC network in a manner that was not possible before.

By having better control over the flow of electricity in a T&D network by FACTS devices, it is expected to lower electricity costs by efficient utilization of existing assets, better outage management, greater operational flexibility, and faster recovery from system disturbances. These devices will also enable greater utilization of economic energy sources (including distributed energy resources) and facilitate competition in the electric supply business.

9.3 FLEXIBLE AC TRANSMISSION

HVDC

The development of FACTS technology has evolved from the early days of High-Voltage DC Transmission (HVDC) where converter stations are used to rectify AC to DC, and an inverter station is used to convert DC back to AC. A converter station is generally composed of 12 or 24 pulse thyristor-controlled electronic bridge circuits. Unlike an AC transmission circuit, a wide variety of control is possible in an HVDC transmission system by controlling the firing angle of thyristors.

There are several HVDC transmission systems operational in the USA. The Pacific Intertie is one example of an HVDC transmission system that is used to bring low-cost hydroelectricity from the Pacific Northwest to California by means of a bipolar ± 500kV HVDC transmission line. HVDC offers several advantages over AC transmission, including lower transmission cost, asynchronous operation of two electrical systems, and support to AC systems when required.

As consumers require better power quality, greater utilization of FACTS devices are also expected in the distribution systems for better voltage regulation, lower harmonics, minimum voltage flicker, and greater reliability of service. The use of FACTS devices in distribution systems will also lead to greater integration of smaller generation systems with lower environmental impact. HVDC LIGHT is another new development for DC power distribution by underground cables to remote locations where there is no local generation available, or as a backup resource for local generation.

Static VAR Compensator (SVC)

Static VAR compensators are reactive power devices that generate or absorb reactive power as required by the electric power system. Passive shunt capacitors and reactors have been used for a long time for reactive power supply, power factor correction, and voltage support, but were limited by their ability to provide continuously variable output to match system requirements. SVC systems are active devices whose output can be controlled continuously to match system requirements very precisely. These devices are used to maintain constant voltage levels, enhance power flow, improve stability, and provide various other improvements listed at the end of this section.

SVC is widely used in the electric power system to increase the transmission capacity of existing lines. The performance of the SVC device to augment power flow in a transmission line is seen in Figure 9.9. In an uncompensated transmission line, power flow is limited by voltage drop on the line, such that any further increase in power transfer results in voltage collapse. If an SVC is connected to the transmission line, the power transmission capacity of the line may be increased up to the full thermal rating of the transmission line conductor as shown in Figure 9.9. This example is given for a 132 kV transmission line having an ACSR Cardinal conductor. Without an SVC device connected on this line, the transmission capacity is limited to 750 MW. When an SVC device is connected, it is possible to utilize the full thermal capacity of the line, up to 1000 MW or more.

The most common types of SVC systems used at present consist of a thyristor-controlled reactor (TCR) and a thyristor-switched capacitor (TSC). The basic elements of a TCR are shown in Figure 9.10, and a TSC is shown in Figure 9.11.

A TCR consists of a reactor in series with a bidirectional thyristor pair. A thyristor is a fast-acting electronic switch, basically a four-layer pnpn device, consisting of anode, cathode, and gate. The thyristor is turned on by applying a short pulse across the gate and cathode, and turned off by applying a reverse voltage across the anode and cathode. The reactive power absorbed by the TCR device is controlled by regulating the current flow through the reactor by directing the firing angle of the thyristor between 90° and 180°. There is full conduction at 90° firing

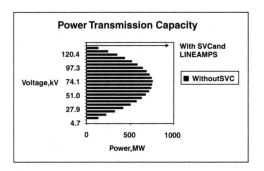

FIGURE 9.9 Increasing power transmission line capacity by FACTS using a Static Var Compensator (SVC).

FIGURE 9.10 Thyristor Controlled Reactor (TCR)

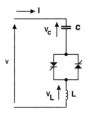

FIGURE 9.11 Thyristor Switched Capacitors (TSC)

angle. Increasing the firing angle decreases current, and conduction is blocked when the angle is 180°. A continuously variable lagging reactor power, Q_L, is made possible by controlling the firing angle of the thyristor given by,

$$Q_\ell = I_\ell (\alpha)^2 X_\ell$$

where,

Q_ℓ = reactive power (lagging)
I_ℓ = reactor current
α = firing angle of thyristor
X_ℓ = inductive reactance

Thyristors having voltage and current ratings up to 5 kV and 4000 A are presently available for switching frequency up to 300 Hz. A basic model of a thyristor-switched capacitor (TSC) system is shown in the Figure 9.11. A TSC is used to deliver leading reactive power, Q_c, by switching on thyristors and is given by,

$$Q_c = v_c^2 / X_c$$

Where,

Q_c = leading reactive power delivered to the system
V_c = capacitor voltage
X_c = capacitive reactance

A combination of TCR and TCS is often connected in parallel to offer a continuously variable SVC system providing leading or lagging reactive power, Q_r, obtained as follows: $Q_r = Q_l - Q_c$. An example of an SVC system having a combination of TCR and TCS in an actual 735 kV substation is shown in Figure 9.12, and the single-line diagram is shown in Figure 9.13.

There are more than 100 SVC installations worldwide operating at voltages up to 500kV with capacity in the range of ± 50 MVA to ± 400 MVA.

FIGURE 9.12 A 735 kV SVC substation, Hydro Quebec Canada.

STATCOM

The development of Gate Turn-Off (GTO) Thyristor technology has recently led to the making of an all solid-state VAR generator called STATCOM (Static Synchronous Compensator). Unlike thyristors, GTOs are capable of turning off by changing their state from conduction to nonconduction by pulsing the gate. At present, GTO

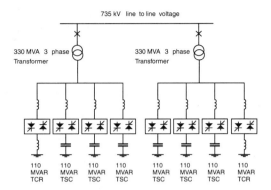

FIGURE 9.13 Single-line diagram of the 735 kV SVC system with TCR and TSC

voltage and current rating is 4 kV and 3000 A, respectively, 1 kHz commutation frequency, and 40 µs speed. This makes the GTO more suitable for faster turning off, requiring less power than a conventional thyristor. Unlike the TCS and TCR devices described previously, STATCOM offers continuously variable reactive power without the use of capacitors or reactors. As a result, the size and cost of static VAR generating equipment is reduced considerably.

The basic operating feature of a STATCOM is similar to the rotating synchronous condenser, but has the advantage of solid-state technology, having no moving parts and with a high degree of reliability. The STATCOM consists of a DC to AC converter circuit, which provides a three-phase output voltage in phase with the AC system voltage as shown in Figure 9.14.

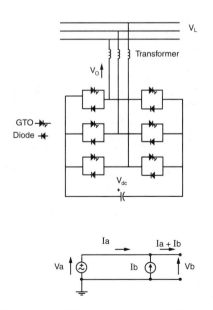

FIGURE 9.14 STATCOM, 6 pulse GTO converter and equivalent circuit

By controlling the magnitude of the converter output voltages, Vo, using GTO, the reactive power delivered by STATCOM can be controlled from full leading to full lagging. Increasing the magnitude of converter output voltages above AC system voltage, V_L, delivers leading VARS to the AC system. A reduction in converter output voltage delivers lagging VARS.

SERIES COMPENSATION

Series compensation has been used for a long time to increase the power transmission capacity of long transmission lines by connecting capacitors in series at suitable locations along the line. Early series compensation devices used mechanical switches to connect capacitor banks in series with a transmission line to reduce line impedance and increase transmission capacity.

Today, controllable series capacitors with GTO thyristors are used to control the degree of reactive power compensation more precisely as a function of network operating conditions. Thus, with a thyristor-controlled series capacitors, (TCSC) it is now possible to control the current flowing through a specific high-voltage line, as well to maximize the ampacity of the line. Some examples of possible series capacitor locations in long transmission lines are shown in Figures 9.15 and 9.16, and a thyristor-controlled series capacitor with a protection device is shown in Figure 9.17.

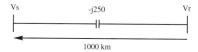

FIGURE 9.15 Series compensated line of Example 1.

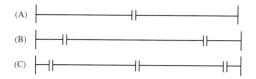

FIGURE 9.16 Location of series capacitors on long transmission lines.

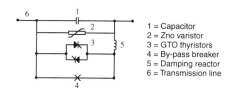

1 = Capacitor
2 = Zno varistor
3 = GTO thyristors
4 = By-pass breaker
5 = Damping reactor
6 = Transmission line

FIGURE 9.17 Controllable series capacitor

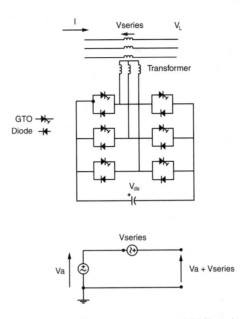

FIGURE 9.18 Static Synchronous Series Compensator (SSSC) and equivalent circuit

Static Synchronous Series Compensators (SSSC) are also developed that offer the faster response time required for the damping of power system oscillations caused by faults and other system disturbances. A GTO circuit is used for SSSC, which is similar to a STATCOM, as shown in Figure 9.18. A complex voltage Vseries is generated by connecting the GTO converter to a transformer connected in series with the transmission line. This complex voltage V series is then added to the AC voltage of the transmission line to control transmission line voltage.

Example 1

A 1000 km 765 kV transmission line delivers natural load. It is proposed to increase the transmission capacity by two times the natural load. A series capacitor having a reactance of -j250 ohm is installed at the middle of the line, as shown in Figure 9.18. Determine the voltage regulation of the line.

Transmission line data

Receiving end voltage = 765 kV
R_{dc} = 0.041 ohm/km
$L = 8.35 \times 10^{-4}$ H/km
$C = 12.78 \times 10^{-9}$ F/km
$G = 0$

Solution

For a 1000 km line, the following long-line equations, derived in Appendix 10.1 at the end of Chapter 10, will be used.

$$\overline{V}_x = \cosh(\overline{\gamma}x)\overline{V}_r + \overline{Z}c \cdot \sinh(\overline{\gamma}x)\overline{I}_r \qquad (9.1)$$

$$\overline{I}_x = \frac{\sinh(\overline{\gamma}x)\overline{V}_r}{\overline{Z}c} + \cosh(\overline{\gamma}x)\overline{I}_r \qquad (9.2)$$

x = distance from receiving end, km
$\overline{V}_x$ = Voltage at a point, x, in the line
$\overline{V}_r$ = Voltage at the receiving end of the line
$\overline{I}_r$ = Current at the receiving end of the line
$\overline{I}_x$ = Current at a point, x, in the line
$\overline{\gamma}$ = propagation constant
Zc = characteristic impedance of the line

The above equation is derived in the appendix at the end of this chapter. The values can be represented by the following matrix equation utilizing the well-known A, B, C, D constants of the line:

$$\begin{bmatrix} \overline{V}s \\ \overline{I}s \end{bmatrix} = \begin{bmatrix} \overline{A} & \overline{B} \\ \overline{C} & \overline{D} \end{bmatrix} \begin{bmatrix} \overline{V}r \\ \overline{I}r \end{bmatrix}$$

The natural load of the line is given by the Surge Impedance Loading (SIL), (See Example 4, Section 10.5).

$$\text{SIL} = 2289 \text{ MVA}$$

Therefore, the load current, Ir, is,

$$\overline{I}r = \frac{\text{SIL}}{\sqrt{3} \cdot 765 \cdot 10^3}$$

$$\overline{I}r = 1728 \text{ A}$$

The propagation constant and the characteristic impedance are obtained from (Example 4, Section 10.5).

$$\overline{\gamma} = 2 \cdot 10^{-5} + j1.2 \cdot 10^{-3}$$

$$\overline{Z}_o = 255.6 - j4.15$$

The sending end voltage Vs is obtained from Section 10.5,

$$\overline{V}_s = \cosh\{(2 \cdot 10^{-5} + j1.23 \cdot 10^{-5}) \cdot 10^3 \cdot 441 \cdot 10^3\}$$
$$+ (255.6 - j4.16) \cdot \sinh\{(2 \cdot 10^{-5} + j1.23 \cdot 10^{-3}) \cdot 10^3\} \cdot 1728$$

$$\overline{V}_s = (156.7 + j424.9) \cdot 10^3$$

$$|\overline{V}_s| = 452.8 \cdot 10^3$$

$$\text{Regulation} = (|V_s| - |V_r|) \cdot \frac{100}{|V_r|}$$

$$= 2.52\%$$

Sending end voltage when load is 2 times SIL,

$$\overline{V}_s = (166.4 + j841.4) \cdot 10^3$$

$$|\overline{V}_s| = 857.7 \cdot 10^3$$

$$\text{Regulation} = (857.7 \cdot 10^3 - 441.7 \cdot 10^3) \cdot \frac{100}{441.7 \cdot 10^3}$$

$$= 94.1\%$$

The sending ending voltage is too high and the regulation is unacceptable.

A series capacitor bank having a total reactance of -j250 ohm is now added at the middle of the line. The ABCD constants of a 500 km section of line are:

$$\overline{A} = \cosh(\gamma \cdot \ell)$$

$$\overline{A} = \overline{D} = \cosh(2 \cdot 10^{-5} + j1.23 \cdot 10^{-3}) \cdot 500$$

$$= 0.816 + j5.79 \cdot 10^{-3}$$

$$\overline{B} = Zo \cdot \sinh(\gamma \cdot \ell)$$

$$\overline{B} = (255.6 - j4.15) \cdot \sinh\{(2 \cdot 10^{-5} + j1.23 \cdot 10^{-3})\} \cdot 500$$

$$\overline{B} = 4.49 + j147.6$$

$$\overline{C} = \frac{\sinh(\gamma \cdot \ell)}{Zo}$$

$$\overline{C} = \frac{\sinh\{(2 \cdot 10^{-5} + j1.23 \cdot 10^{-3})\} \cdot 500}{(255.6 - j4.15)}.$$

$$\overline{C} = -4.77 \cdot 10^{-6} + j2.26 \cdot 10^{-3}$$

The ABCD constants of the series capacitor are:

$$\overline{A}' = 1$$

$$\overline{B}' = j250$$

$$\overline{C}' = 0$$

$$\overline{C}' = 1$$

The modified ABCD constants of the series compensated transmission line are:

$$\begin{bmatrix} \overline{A}'' & \overline{B}'' \\ \overline{C}'' & \overline{D}'' \end{bmatrix} = \begin{bmatrix} \overline{A} & \overline{B} \\ \overline{C} & \overline{D} \end{bmatrix} \begin{bmatrix} \overline{A}' & \overline{B}' \\ \overline{C}' & \overline{D}' \end{bmatrix} \begin{bmatrix} \overline{A} & \overline{B} \\ \overline{C} & \overline{D} \end{bmatrix}$$

$$= \begin{bmatrix} (0,816 + j5.79) & (4.49 + j147.6) \\ -(4.77 \cdot 10^{-6} - j2.26 \cdot 10^{-3}) & (0.816 + j5.79) \end{bmatrix} \begin{bmatrix} 1 & -j250 \\ 0 & 1 \end{bmatrix}$$

$$\begin{bmatrix} (0.816 + j5.79) & (4.49 + j147.6) \\ -(4.77 \cdot 10^{-6} - j2.26 \cdot 10^{-3}) & (0.816 + j5.79) \end{bmatrix}$$

$$= \begin{bmatrix} (0.794 + j0.023) & (7.993 + j74.515) \\ -(3.935 \cdot 10^5 - j4.966 \cdot 10^{-3}) & (0.794 + j0.023) \end{bmatrix}$$

$$\overline{V}_s = \overline{A}'' \cdot \overline{V}_r + \overline{B}'' \cdot \overline{I}_r$$

$$\overline{V}_s = (.794 + j0.023) \cdot 441.7 \cdot 10^3 + (7.993 + j74.515) \cdot 3455$$

$$\overline{V}_s = 3.782 \cdot 10^5 + j2.677 \cdot 10^5$$

$$\overline{V}_s = 463.4 \cdot 10^3$$

$$\text{Regulation} = (463.4 \cdot 10^3 - 441.7 \cdot 10^3) \cdot \frac{100}{441.7 \cdot 10^3}$$

$$= 4.91\%$$

The above sending end voltage and regulation is acceptable.

The above example shows that by adding a series capacitor at the middle of a line, the maximum power delivered by the line is increased from 2289 MVA natural load to 4578 MVA, which is twice the natural load of the line. Having a GTO thyristor-controlled series capacitor as shown in Figure 9.17 allows control of the current passing through the capacitor and, hence, the reactance, by modifying the A,B,C,D constants of the line according to network operating conditions.

Unified Power Flow Controller (UPFC)

The UPFC is a new development that offers multiple compensation functions by providing independent control of the following transmission line parameters:

- Bus voltage
- Active power
- Reactive power

The UPFC is made of two GTO-based converters connected by a common DC link. They can also operate independently with Converter 1 acting as a STATCOM and Converter 2 like an SSSC. A block diagram of the UPFC converter, control domain, and equivalent circuit is shown in Figure 9.19. Recent UPFC installations worldwide are in the range ± 100 to ± 200 MVA at transmission voltage levels. The AEP* installation (1997–1998) is rated at ± 160 MVA and is located on a 138 kV transmission line.

Custom Power

Custom Power concept covers a number of power electronics devices suitable for connection at the distribution system level. These devices typically have a range from 1 to 10 MVA and may be connected at customer point of connection to provide better voltage regulation, and prevent plant shutdown during adverse voltage con-

---

* World's first Unified Power Flow Controller on the AEP system, B.A. Rena, et al. Cigré 1998.

New Methods of Increasing Transmission Capacity

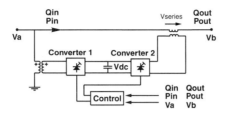

FIGURE 9.19a Unified Power Flow Controller (UPFC), Converter schematic

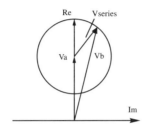

FIGURE 9.19b Unified Power Flow Controller (UPFC), Control domain

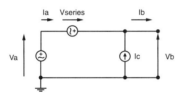

FIGURE 9.19c Unified Power Flow Controller (UPFC), Equivalent circuit1

ditions such as voltage dips and power surges. A distribution system STATCOM is used at the distribution level for power factor correction as more current is passed through the distribution circuit, and for protecting the distribution system from the effects of non-linear loads and flicker reduction applications.

Power electronics devices also enable greater integration of smaller generating systems located closer to the loads. Small generation technologies include micro-turbines, fuel cells, wind turbines, and photovoltaic and other renewable energy sources. In this scenario, the installation of power electronic devices offers faster connection to the main supply in the event of a failure, damping of oscillations, and smoothing out the voltage delivered by the small generation systems.

SMES

The Superconducting Magnetic Energy Storage (SMES) device is another FACTS device (Feak, 1997), (Borgard, L., 1999), (Buckles et al. 2000) that is used for voltage support, damping of power system oscillations, and improvement of power system stability.

During normal operating conditions, energy is stored in a superconducting coil, 6, comprise the SMES device. The stored energy released to the network when there is a disturbance is used to provide voltage support, as well as for damping power system oscillations. A simplified SMES device is shown in Figure 9.20. The two GTO thyristor converters, 4 and 5, are required for conversion of DC power supplied by the superconducting coil to 60 Hz ac. The AC output voltage of the converters is stepped up to the transmission line voltage by a three-winding transformer, 3.

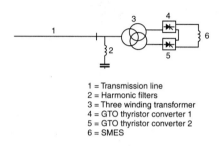

1 = Transmission line
2 = Harmonic filters
3 = Three winding transformer
4 = GTO thyristor converter 1
5 = GTO thyristor converter 2
6 = SMES

FIGURE 9.20 SMES device connected to a high voltage transmission line

LIST OF FACTS APPLICATIONS

A number of FACTS device systems are in use worldwide to solve various power system problems:

- Voltage control
- Load balancing
- Increase active power transmission capacity of existing and future lines
- Increase transient stability margin
- Increase damping of power oscillations caused by a disturbance in the power system
- Facilitate greater use of dispersed, small, and distributed energy sources
- Reduce temporary overvoltages
- Damp subsynchronous resonance
- Provide reactive power to AC-DC converters
- FACTS devices for distribution system and custom power

MANUFACTURERS

There are several manufacturers of FACTS devices in the US and worldwide. Following is a list of important manufacturers.

ABB
GE Power Electronics Division
GEC Alsthom T&D Power Electronics System
Hitachi
SIEMENS
Westinghouse Electric Company

FUTURE RESEARCH AND DEVELOPMENT

The FACTS devices used at present are mostly independent units and are not yet widely integrated with other FACTS devices in the network. Further R&D is required to develop integrated control of several FACTS devices by communication between devices in order to exercise much wider and distributed control of the power system.

For HV and EHV network applications, there is a need to increase the current and voltage rating of power semiconductor devices as well as to increase the switching frequency for higher commutation speed and fewer losses.

This chapter mainly focused on FACTS technology for transmission systems, with a brief discussion on distribution system applications. As consumers require better power quality, greater utilization of FACTS devices is expected for better voltage regulation, lower harmonics, minimum voltage flicker, and greater reliability of service. The use of FACTS devices in distribution systems will also lead to greater integration of smaller generation systems with lower environmental impact.

Another interesting application of FACTS for distribution systems is HVDC LIGHT. This technology is providing competition for distributed generation systems. HVDC LIGHT applies FACTS technology for low voltage DC transmission by underground cables. Low-voltage DC converters are used to tap off connections from HV and EHV AC lines and convert AC to low-voltage DC. The low-voltage DC power supply is then distributed by DC cables to remote areas. Since DC voltage is used for distribution it does not have the problem of the high capacitive reactance of AC underground cables. Therefore, it appears that low-voltage underground DC transmission can supply power for longer distances to remote areas. Underground DC is also more acceptable for environmental reasons.

9.4 CHAPTER SUMMARY

Power electronics semiconductor device technology for FACTS has evolved considerably during the last decade, and several applications of the different technologies are presented in this chapter. For power semiconductor devices, the main requirements are fast turn-on and turn-off times with high current and voltage ratings. At the present time these devices are developed with voltage ratings up to 10 kV, and maximum current ratings in the range of 1000 A to 3000 A. Superconducting Magnetic Energy Storage (SMES) systems are developed for application in all voltage levels for faster response time. SMES provides continuous VAR support and maintains the stability of the interconnected transmission grid.

The development and application of FACTS devices are largely dependent upon semiconductor switching devices. The IGBT device appears to be promising for high-voltage applications requiring fast turn-on and turn-off times and having high switching frequency. GTO thyristors are also available with sufficient voltage and current rating and fast switching capability. Further development of semiconductor power electronic devices is required to obtain greater current and voltage ratings with faster response time and higher switching frequency.

A wide variety of FACTS devices are presented in this chapter, which includes both static series and shunt-compensated devices, and HVDC LIGHT for low-voltage distribution to remote locations. Thyristor controlled series capacitors are particularly useful for increasing the power transmission capacity of long lines without causing synchronous resonance problems associated with generators. The development and application of FACTS technology is essential for the control of current and voltage in transmission and distribution circuits. This will lead to increased reliability of service, the lowering of electricity cost, and better power quality. When lines are operated close to thermal ratings, these new FACTS devices will respond quickly in the event of a system disturbance by providing voltage support, and will maintain generator stability. It is expected that these developments will enable maximum utilization of existing transmission and distribution system resources by making use of the line ampacity system.

10 Applications

10.1 INTRODUCTION

The application of the powerline ampacity system to the economic operation of transmission network, power system stability, and transmission planning is presented in this chapter. The economic load dispatch problem and the optimal power flow problem are presented by consideration of variable transmission line ratings. Results are presented by example of a hypothetical utility generation and transmission network, which clearly demonstrates the economic benefit of operating a power system by utilizing a dynamic powerline ampacity system.

For transmission planning purposes, it is necessary to clearly understand the factors affecting the choice of conductors for overhead transmission line applications. For this purpose, a formulation of the economic sizing of conductors is presented, followed by a study of economic conductor temperatures. Based on the above considerations, the cost of adding a new line in the region of the San Francisco Bay area, for example, is compared to the increased cost of losses due to higher transmission line current. It is shown that the construction of a new line can be deferred by at least ten years by adopting a dynamic line rating system in this region.

For future transmission system planning, due consideration must be given to utilizing renewable energy sources that are generally located far away from major metropolitan areas and industrial load centers. Extra High Voltage (EHV) transmission lines are required to transport electric energy economically over long distances. For this reason, some important factors affecting long-distance transmission, mainly from the point of view of transmission line ampacity, are discussed.

The maximum power transmission capacity of certain overhead powerlines is sometimes limited by the stability of generators connected to transmission lines. Therefore, the effect of increasing line ampacity on generator stability is presented. The steady, dynamic, and transient stabilities of generators are then related to the steady, dynamic, and transient line ratings.

10.2 ECONOMIC OPERATION

In an earlier study, the author performed economic evaluation of an interconnected electricity generation and transmission system by a dynamic line rating system (Hall, Deb, 1988a). In this study, actual PG&E transmission and generation system costs were presented to show cost savings by dynamic line ratings. We presented a case study without performing a formal network optimization analysis. Therefore, the present work on generation and transmission system cost optimization by dynamic line rating is an extension of previous work on this subject.

10.2.1 FORMULATION OF THE OPTIMIZATION PROBLEM

It is required to minimize the total cost of electricity production, including generation and transmission cost for a given load demand and operating system constraints. In an interconnected electric utility system comprising dispersed generation sources, the problem is how to allocate the required load demand among the available generation units on an hour-by-hour basis. This is carried out by minimizing an objective function, F, that represents the total cost of producing electricity.

The objective function is given as,

$$F = \sum_{i=1}^{n} \left(a_i + b_i \cdot P_i + c_i \cdot P_i^2 \right) \quad (10.1)$$

where,

$a_i, b_i, c_i,$ = cost coefficients of the i-th generator
n = number of generator units
P_i = power output of the i-th generator

The objective function (10.1) is subject to the following constraints:

Power Balance

This requires that the sum of all power generated be equal to the sum of demand and transmission loss.

$$\sum_{i=1}^{n} P_i - P^D - P^L = 0 \quad (10.2)$$

P^D = demand
P^L = transmission loss given by,

$$P^L = \sum_{i=1}^{n} B_i P_i^2 \quad (10.3)$$

B_i is the transmission loss coefficient.

Nodal Balance of Active and Reactive Powers (Load-Flow Equations)

Active Power

$$P_i = \sum_{k=1}^{n} |V_i||V_k| \left[g_{i,k} \cos(\theta_i - \theta_k) + b_{i,k} \sin(\theta_i - \theta_k) \right] \quad i = 1,\ldots n \text{ nodes} \quad (10.4)$$

Applications

Reactive Power

$$Q_i = \sum_{k=1}^{n} |V_i||V_k|\left[g_{i,k}\text{Cos}(\theta_i - \theta_k) + b_{i,k}\text{Sin}(\theta_i - \theta_k)\right] \quad i = 1,\ldots n \text{ nodes}$$

$$\theta \cong \angle V_i \tag{10.5}$$

V = node voltage

Transmission Line Capacity Limits (Line Ratings)

Line current between nodes i and k:

$$I_{i,k} = \frac{|V_i \angle \theta_i - V_k \angle \theta_k|}{\bar{z}_{i,k}} \leq I_{i,k}(\text{Max})$$

$$\bar{z}_{i,k} = \frac{1}{g_{i,k} + jb_{i,k}} \tag{10.6}$$

$I_{i,k}(\text{Max})$ = Dynamic line rating

Generator Capacity Limits

The active (P) and reactive (Q) power output of the i-th generator should be between minimum and maximum generation limits:

$$P_{\min,i} \leq P_i < P_{\max,i} \quad i = 1,\ldots.\text{Ng} \tag{10.7}$$

$$Q_{\min,i} < Q_i \leq Q_{\max,i} \quad i = 1,\ldots.\text{Ng} \tag{10.8}$$

Ng = Number of generator nodes

Bus Voltage Limits

$$V_{\min,i} \leq V_i \leq V_{\max,i} \quad i = 1,\ldots N \text{ nodes} \tag{10.9}$$

Transformer Tap Limits

$$VT_{\min,i} \leq VT_i \leq VT_{\max,i} \quad i = 1,\ldots\text{Ntap} \tag{10.10}$$

The above equations describe the optimal power-flow problem. Additional constraints may be added to consider emission levels in fossil fuel-based generation units. A solution to the above nonlinear optimization problem is available in a modern power system analysis textbook (Bergen, A., 1986). More recently, artificial neural network (Lee, K.Y. et al., 1998), (Yalcinov, Short, 1998) and genetic algorithm GA (Wong, Yuryevich, 1998) solutions have also been proposed. A simplified solution of the above problem is presented in Table 2.3 and Figures 10.3 and 10.4 for a static and dynamic line rating system. The object of this study is to show the significance of dynamic transmission line ratings in the economic operation of an interconnected power system having diverse generation sources.

10.2.2 ELECTRICITY GENERATION COST SAVING IN INTERCONNECTED TRANSMISSION NETWORK

To illustrate the main concepts presented in the previous section, a hypothetical electric utility generation and transmission system was created as shown in Figures 10.3 and 10.4. The transmission network receives power from five generation sources (G1-G5). The generation sources G1 and G3 are hydroelectric, and all other generation sources are thermal. The transmission network is comprised of five 230 kV double circuit lines and the network data is shown in the Table 10.1. An ACSR Cardinal conductor is used in each phase of the line. The dimensions of the 230kV double circuit line are given in Figure 10.1. The transmission line electrical π equivalent model shown in Figure 10.2 was used to obtain network load-flow solutions.

TABLE 10.1
230 kV Transmission network data with ACSR Cardinal Conductor

From Bus	To Bus	Line #	Impedance $Z_1 = R_1 + jX_1$	Half-Line Charging Capacitance $\dfrac{Y_s}{2} = \dfrac{j\omega C}{2}$
1	2	1	0.02 + j0.13	j0.03
1	5	2	0.05 + j0.32	j0.025
2	3	3	0.016 + j0.10	j0.02
3	4	4	0.02 + j0.13	j0.02
4	5	5	0.016 + j0.10	j0.015

A comparison of electricity generation costs is made in the following sections using static and dynamic line rating systems. It is shown that substantial economy in electricity generation cost can be achieved by adopting a dynamic line rating system.

Electricity Generation Cost Using Static Line Rating System

Electric power companies generally follow a static line rating system by assuming constant line capacity based on conservative assumptions of the meteorological parameters that affect line capacity. Many electric power companies have seasonal line ratings for summer and winter. There are several reasons for operating lines based on static line ratings. For example, transmission lines are operated more easily

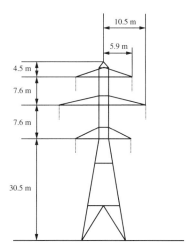

FIGURE 10.1 230 kV Double Circuit Transmission Line Tower.

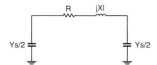

FIGURE 10.2 Transmission Line Electrical PI model.

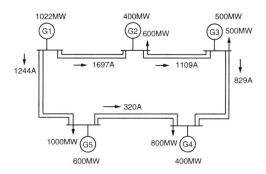

FIGURE 10.3 Electricity generation and transmission line current using static line rating.

by following a static line rating system. The transmission line protection scheme does not have to be changed because line current limits are held constant. Also, close monitoring of weather conditions all along the transmission line route, as well as monitoring transmission line conductor temperatures, conductor tension, and other parameters are not required.

In the example of Figure 10.3, the static line rating of each transmission line phase conductor is assumed to be 849 A for normal operating conditions. An ACSR

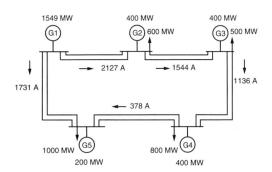

FIGURE 10.4 Electricity generation and line current using dynamic line rating.

Cardinal conductor is used in all lines. Based on the above assumptions, the optimum power-flow solution of the network is shown in Figure 10.3 and Table 10.2. In Figure 10.3 it is seen that the ampacity of Line 1 has reached its maximum static rating (2 × 848 = 1996 A), and the total cost of electricity generation for meeting the demand for one hour is $ 48,877.32.

Electricity Generation Cost Using LINEAMPS Rating

Due to environmental and economic considerations, many electric power companies have now started to use a system of rating transmission lines that is variable, depending upon actual transmission line operating conditions (Soto et al., 1998), (Wook et al., 1997), (Steeley et al., 1991). As stated in the previous chapters, the variable system of line ratings is commonly known as "dynamic line rating system," where transmission line ampacity is adapted to actual and forecast weather conditions. In this section the economic benefits of a dynamic line rating system are demonstrated by an example of the transmission network considered in the previous section.

In the optimum power-flow solution of Figure 10.3, the capacity of Line 1 reached its maximum static ampacity, which limited further addition of cheaper hydroelectricity available from the generator (G1). It is shown that by utilizing a dynamic line rating system the network is able to utilize a greater percentage of hydroelectric generation from the generation source (G1).

The optimum power-flow solution achieved by dynamic line rating is presented in Figure 10.4. As seen in this figure the network was able to receive 1549 MW of cheaper hydroelectric energy from generator source (G1) due to the higher transmission line ampacity (2 x 1063.5 = 2127 A) offered by LINEAMPS. A similar result was obtained in a previous study (Hall, Deb, 1988a).*

Cost Savings by LINEAMPS

The power flow solution presented in Table 10.3 utilizes dynamic line rating, which has resulted in an electricity cost saving of $ 4020.87 for one hour. Such favorable conditions exist often during the life of a transmission line when it is required to

---

* Assuming typical electricity generation cost at PG&E.

Applications

transfer power greater than static line rating. For a more realistic example, the network of Figure 10.3 may be considered a simplified transmission network of the San Francisco region. The generation source (G1) may represent the cheaper source of hydroelectric energy from the Pacific Northwest supplying power to San Francisco Bay area. Similarly, the generation source (G3) could well represent hydroelectric energy supplied from the Sierras. Sometimes there is surplus electric energy available from these sources due to greater than normal rainfall or snow in these areas. Using LINEAMPS ratings, these cheaper sources of electricity can be utilized in the Bay Area with substantial cost savings as shown in Figure 10.4 and Table 10.3.

TABLE 10.2
Electricity Production Cost: Static Line Rating

Bus Number	Generation MW	Load MW	Generation Cost $/MWh	Generation Cost $/hr
1	1022.07	0	10.30	10527.32
2	400	600	24.20	9680.00
3	500	500	12.50	6250.00
4	400	800	25.30	10120.00
5	600	1000	20.50	12300.00
			Total Generation Cost, $/hr	48877.32

TABLE 10.3
Electricity Production Cost: Dynamic Line Rating

Bus Number	Generation MW	Load MW	Generation Cost $/MWh	Generation Cost $/hr
1	1549.17	0	10.30	15956.45
2	400	600	24.20	9680.00
3	400	500	12.50	5000.00
4	400	800	25.30	10120.00
5	200	1000	20.50	4100.00
			Total Generation Cost, $/hr	44856.45

In addition to operational cost savings by facilitating economy energy transfer as mentioned above, it is also possible to save the capital investment required for the construction of new lines or the reconductoring of existing lines. By following the economic analysis presented in Section 10.4, it is shown in Figure 10.5 that capital investment for new line construction can be deferred for at least ten years by increasing the capacity of existing lines for different overload conditions.

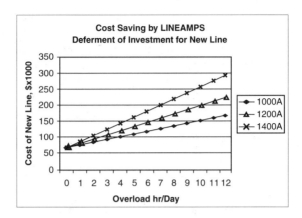

FIGURE 10.5 Cost of adding a new line is compared to the capitalized cost of increased transmission line losses for a range of overload conditions. The analysis is carried out for a period of ten years. For example, if we consider 1000 A overload for 2 hr/Day we can see from this graph that it is more economical to allow the overload current than to build a new line having cost greater than $75,000 /km. The above example is for a 230 kV line with ACSR Cardinal conductor having a static line rating of 849 A.

10.3 STABILITY

The maximum power transmission capacity of certain overhead powerlines are limited by voltage drop and the stability of generators connected to transmission lines. In Chapter 9 the various methods of increasing transmission capacity by the installation of modern power electronics devices were presented. In this section the method of calculation of generation stability when supplying power over a transmission line is presented briefly. The following types of stability (Nasar, Trutt, 1998) problems are discussed:

- Steady-state stability
- Dynamic stability
- Transient stability

We are concerned with the steady-state stability of a generator when the power transferred over a line is increased slowly. The steady-state stability limit of a single generator supplying a load through a transmission line is given by,

$$Ps = \frac{|Vs||Vr|\sin(\delta)}{X} \qquad (10.11)$$

where,

Ps = steady state power transfer
Vs = sending end voltage
Vr = receiving end voltage

Applications

X = line reactance
δ = generator rotor angle

Maximum steady-state power, Pmax, is obtained when the rotor angle δ = 90°. Increasing δ > 90° will result in lower power transfer and, ultimately, loss of steady-state stability as Ps approaches zero. From Equation 10.11, one possible means of increasing steady-state stability is to add capacitors in series with the line to lower the reactance, X, of the line. This was discussed in Chapter 9 in Series Compensation.

Example 10.1

A generator is supplying a load through a 50 km 230 kV double circuit line. Calculate the maximum power transfer capability of the line. The transmission line data is given below.

Conductor = ACSR Cardinal
Reactance of line = 0.5 ohm/km
Transmission line sending and receiving end voltage magnitude = 230kV

Solution

Selecting generator base MVA = 1000 MVA, 3 phase
Selecting transmission line base kV = 230 kV

Transmission line base impedance = $\dfrac{230^2}{1000}$ = 52.9 ohm

Line impedance = $\dfrac{50 \cdot 0.5}{52.9}$ = 0.47 pu

Pmax $\dfrac{1}{0.47}$ sin 90
Pmax = 2.12 pu
Pmax = 2.12x1000 = 2120 MVA

10.3.1 Dynamic Stability

Dynamic stability is concerned with generator oscillations due to step changes in load or other small disturbances. Small changes in generator output due to load variations result in generator rotor oscillations. If oscillations increase, in time the system becomes unstable. The system is dynamically stable if the oscillations diminish with time, and the generators return to a stable state. Generally, the dynamic condition oscillations remain for several seconds until steady-state conditions are reached.

The differential equation governing rotor motion, $\Delta\delta$, with respect to time, t, due to a small increase in power, ΔP, is obtained by (Saadat et al., 1998):

$$\frac{H}{\pi f_0}\frac{d^2\Delta\delta}{dt^2} + D\frac{d\Delta\delta}{dt} + P_s\Delta\delta = \Delta P \tag{10.12}$$

$$\frac{d^2\Delta\delta}{dt^2} + 2\zeta\omega_n\frac{d\Delta\delta}{dt} + \omega_n^2\Delta\delta = \frac{\pi f_0 \Delta P}{H} \tag{10.13}$$

where,

$\Delta\delta$ = small deviation in power angle from initial operating point δ_0
D = damping constant
H = Inertia constant
f_0 = frequency
ω_n = natural frequency of oscillation
ζ = damping ratio given by,

$$\zeta = \frac{D}{2}\sqrt{\frac{\pi \cdot f_0}{H \cdot P_s}} \tag{10.14}$$

The above differential equation is obtained by linearization of the swing equation (Bergen 1986) and is applicable for small disturbances only.

The solution of the above differential equation is (Saadat et al 1998):

$$\Delta\delta = \frac{\pi \cdot f_0 \cdot \Delta P}{H \cdot \omega_n^2}\left[1 - \frac{1}{\sqrt{1-\zeta^2}} \cdot e^{-\zeta\omega_n t}\sin(\omega_d t + \theta)\right] \tag{10.15}$$

Example 10.2

The transmission line of Example 10.1 delivers a load of 1000 MVA under steady-state conditions. Show that the system will remain stable if the load is suddenly increased to 1200 MVA. Assume generator frequency is 60 Hz during normal operating conditions, and the inertia constant is H = 6 pu. The damping constant is D = 0.138 pu.

Solution

Initial operating angle $\delta_0 = \sin^{-1}\left(\frac{Pm}{P\max}\right)$

$$\delta_0 = \sin^{-1}\left(\frac{0.6}{2.128}\right) = 16.38°$$

Ps = P max $\cdot \cos(\delta_0)$ = 2.04 pu

$$\omega_n = \sqrt{\pi \frac{60}{H \cdot Ps}}$$

$$\zeta = \frac{D \cdot \sqrt{\frac{\pi \cdot 60}{H \cdot Ps}}}{2} = 0.271$$

$$\omega_d = \omega_n \cdot \sqrt{1-\zeta^2} = 7.709$$

$$\theta = \cos^{-1}(\zeta)$$

$$\delta(t) = 0.286 + \pi \cdot 60 \cdot \frac{0.2}{6 \cdot 8^2}\left[1 - e^{(-0.271 \cdot 8 \cdot t)} \cdot \frac{\sin(7.707 \cdot t + 1.29)}{\sqrt{1-0.271^2}}\right]$$

The solution of the above equation for a period of 3 s is shown in Figure 10.6. As seen in the figure, the system returns to a stable state after a sudden increase in transmission line load from 1000 MVA to 1200 MVA.

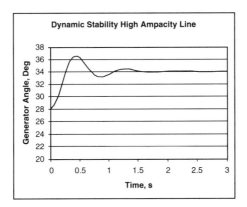

FIGURE 10.6 Generator rotor angle oscillations due to sudden increase in transmission line load current from 1000 MVA to 1200 MVA.

10.3.2 Transient Stability

Transient stability is concerned with generator oscillations due to sudden changes in power transfer levels caused by large disturbances which are due to short-circuit, large scale load-shedding, or generator or transmission line outage. Since we are dealing with large disturbances, linearization of the swing equation is not possible. Numerical solution of the nonlinear differential equation is obtained by Euler's or the Runge Kutte method. A simplified swing equation neglecting damping for transient stability studies is given by,

$$\frac{d^2\delta}{dt^2} - \frac{\pi \cdot f_0 \cdot \Delta P}{H} = 0 \qquad (10.16)$$

Following Euler's method, we obtain the change in angle, $\Delta\delta_n$, during a small interval, Δt:

$$\Delta\delta_n = \delta_n - \delta_{n-1}$$

$$\Delta\delta_n - \Delta\delta_{n-1} = \Delta t \cdot \left(\omega_{n-1/2} - \omega_{n-3/2}\right)$$

$$= \Delta t \cdot \Delta t \frac{\pi \cdot f_0 \cdot \Delta P_{n-1}}{H}$$

$$\Delta\delta_n = \Delta\delta_{n-1} + \frac{\pi \cdot f_0 \cdot \Delta P_{n-1} \cdot (\Delta t)^2}{H}$$

Example 10.3

The transmission line of Example 10.2 is delivering 1200 MVA through both circuits when a short-circuit occurs on one transmission line that lowers the power to 400 MVA. The fault is cleared in 0.125s. The power delivered by the line is 1200 MVA after the fault. Determine if the system will remain stable and attain a steady-state operating condition.

Solution

Initial operating angle,

$$\delta_0 = \sin^{-1}\left(\frac{P_o}{P\max}\right) = \sin^{-1}\left(\frac{1.2}{2.04}\right)$$

$$= 36°$$

The accelerating power is $\Delta P = 0$ before the fault.
Acceleration power, ΔP, after the fault is:

$$\Delta P = P_0 - 0.4 \sin \delta_0$$

$$= 0.96$$

Starting with $t = 0$ and $\delta = 36°$ and time interval $\Delta t = 0.05$ we find,

$$\Delta\delta_n = \Delta\delta_{n+1} + \frac{180 \cdot 60 \cdot (\Delta t)^2 \cdot \Delta P_{n-1}}{H \cdot P_0}$$

Applications

During fault we obtain the change in rotor $\Delta\delta$ angle during a small interval, Δt, by,

$$\Delta\delta_n = \Delta\delta_{n-1} + \frac{180 \cdot 60 \cdot (\Delta t)^2 \cdot [P_0 - 0.4\sin(\delta_{n-1})]}{H \cdot P_0}$$

and after the fault,

$$\Delta\delta_n = \Delta\delta_{n-1} + \frac{180 \cdot 60 \cdot (\Delta t)^2 \cdot [P_0 - 1.2\sin(\delta_{n-1})]}{H \cdot P_0}$$

The above equation is solved for a period of 2 s. The swing curve is stable as shown in Figure 10 7.

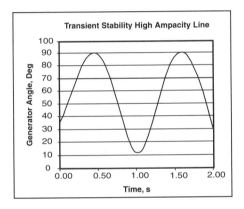

FIGURE 10.7 Swing curve for a transmission line fault cleared in 0.125 s.

10.4 TRANSMISSION PLANNING

In the previous section, the application of the powerline ampacity system to power system operations was presented. In this section, applications to transmission system planning and design are discussed. This study includes transmission system cost analysis, optimum sizing of transmission line conductors, and the evaluation of optimum conductor temperatures. The following factors are considered:

- Capital cost of line
- Cost of capital (interest rate)
- Cost of energy, $/MWH
- Load factor
- Conductivity of conductor material

Specific Transmission Cost

The economic evaluation of transmission lines is carried out by comparing the specific transmission cost (Vp) of different alternatives. The specific transmission cost is defined as the cost per MVA/Km of power delivered by the line given as follows (Hall, Deb, 1988a):

$$Vp = \frac{Wp}{MVA} \qquad (10.17)$$

Wp = Present worth of line = Cp + Co
Cp = Capital cost of line
Co = Capitalized cost of line operation

Capital Cost of Line

For estimation purposes the capital cost of line, Cp, may be obtained by,

$$Cp = a_1 + a_2 S + a_3 V \qquad (10.18)$$

Where, a_1, a_2, a_3 are the coefficients of the line cost model obtained by statistical fitting of historical data of line costs at different transmission voltage (V) and conductor size (S).

Capitalized Cost of Line Operation

The capitalized cost of line operations (Co) includes the cost of losses (Cl) and the cost of line maintenance (Cm). It is calculated as follows,

$$Co = k(Cl_j + Cm_j) \; j = 1,2\ldots n \text{ years} \qquad (10.19)$$

k = capitalization factor = $\sum_{j=1}^{n}(1+i)^{-j}$

n = life of the line, years
i = interest rate

The annual cost of line losses C_1 is obtained by,

$$C_1 = \frac{3 \cdot I^2 \cdot R_{ac} \cdot L_s \cdot d \cdot 8760}{S} \qquad (10.20)$$

I = conductor current, A
R_{ac} = ac resistance of conductor, ohm/km
L_s = load loss factor
d = cost of energy, $/MWH

Applications

The load loss factor L_s is related to the load factor L_f by,

$$L_s = k_1 \cdot L_f + k_2 \cdot L_f^2 \qquad (10.21)$$

Where, k_1, k_2 are constants (Hall, Deb, 1988a).
The load factor L_f is defined as,

$$L_f = \frac{\text{Energy supplied by the line in a year}}{\text{Maximum demand} \times 8760} \qquad (10.22)$$

Optimum Size of Conductor

When planning a new transmission line, it is required to select the optimum size of conductor for a given maximum power transfer. The optimum size of the conductor is obtained by minimizing the present worth (Wp) of the total transmission cost as follows:

$$\text{Min}(Wp) = \frac{dWp}{dS} = 0 \qquad (10.23)$$

From equations 10.17–10.23 we obtain,

$$Wp = k\left[\frac{3 \cdot I^2 \cdot r \cdot l \cdot L_s \cdot d \cdot 8760}{S} + Cm\right] + (a_1 + a_2 S + a_3 V) \qquad (10.24)$$

With the help of Equation 10.24, we can perform transmission cost evaluation studies with alternative conductor designs for transmission planning purposes as shown in the Table 10.4 adapted from (Anand et al. 1985).

$$R_{ac} = \frac{r \cdot l}{S}$$

r = resistivity of conductor, ohm·mm²· m⁻¹
l = length of line = 1 km
S = sectional area of conductor, mm²

Differentiating Wp with respect to S and setting $\frac{dWp}{dS} = 0$, we obtain the optimum size of conductor,

$$S_{optimum} = I \sqrt{\frac{3 \cdot k \cdot L_s \cdot d \cdot r \cdot 8760}{a_2}} \qquad (10.25)$$

and the optimum current density,

$$J_{optimum} = \sqrt{\frac{a_2}{3 \cdot k \cdot L_s \cdot d \cdot r \cdot 8760}} \qquad (10.26)$$

Equation 10.26 shows that optimum current density depends upon the factor a_2, which represents that portion of the capital cost of line that depends upon conductor size, S; interest rate, ($k \propto i$); Load factor, ($L_f \propto L_s$); energy cost, d; and the resistivity of conductor material, r. The value of r is important — as we approach superconductivity, the optimum current density, J, will become very high.

TABLE 10.4

Line Data	ACSR 54/7	AAAC(1) 61	AAAC(2) 61	ACAR 54/7	ACSR/AS20 54/7	Compact 54/7
MVA capacity	780	790	810	815	805	800
Economic span, m	425	475	425	400	425	425
Tension, kN	43	48	38	36	43	43
Loss[1]	1.0	0.99	0.97	1.06	0.98	0.98
Line cost[1]	1.0	0.95	0.96	0.97	0.97	0.97
Transmission cost[1]	1.0	0.97	0.96	1.01	0.97	0.97

[1]Cost with respect to ACSR
Conductor diameter = 31 mm
AAAC (1) = 53% IACS
AAAC (2) = 56% IACS

10.5 LONG-DISTANCE TRANSMISSION

Extra High Voltage (EHV) transmission lines are used for the transportation of electric energy over long distances economically. At the present time the highest transmission line voltage in North America is 765 kV; there are some experimental lines capable of reaching voltages up to 1100 kV but they are not in operation. Transmission line voltage up to 800 kV is operational in many countries, and a 1150 kV EHV AC line is operating in Russia (Alexandrov et al. 1998).

For long-distance transmission line analysis, a lumped parameter equivalent of a line is no longer accurate, and a distributed parameter representation of the line is used for transmission line analysis. The following distributed parameter equations of the transmission line are used for the analysis of transmission line voltage and current along the length of the line:

$$\overline{V}_x = \cosh(\overline{\gamma}x)\overline{V}_r + \overline{Z}c \cdot \sinh(\overline{\gamma}x)\overline{I}_r \qquad (10.27)$$

Applications

$$\bar{I}_x = \frac{\sinh(\bar{\gamma}x)\bar{V}_r}{\bar{Z}c} + \cosh(\bar{\gamma}x)\bar{I}_r \qquad (10.28)$$

x = distance from receiving end, km
$\bar{V}_x$ = Voltage at a point x in line
$\bar{V}_r$ = Voltage at the receiving end of line
$\bar{I}_r$ = Current at the receiving end of line
$\bar{I}_x$ = Current at a point x in line
$\bar{\gamma}$ = propagation constant
Zc = characteristic impedance of the line

The above equation is derived from Appendix 10 at the end of this chapter. It can be represented by the following matrix equation utilizing the well known A, B, C, D constants of the line.

$$\begin{bmatrix} Vs \\ Is \end{bmatrix} = \begin{bmatrix} A & B \\ C & D \end{bmatrix} \begin{bmatrix} Vr \\ Ir \end{bmatrix} \qquad (10.29)$$

The following example will illustrate some interesting features of long-distance transmission.

Example 10.4

It is proposed to supply large amounts of cheap hydroelectricity by Extra High Voltage (EHV) transmission line from a location 2500 km away from the load center. The transmission line voltage is 765 kV AC. Find the following:

1. Surge Impedance Loading for this line
2. Line ampacity and maximum power transmission capacity
3. Transmission line current as a function of line distance for power transmission equal to 0.5 SIL, 1 SIL, and 2 SIL.
4. Transmission line voltage as a function of line distance for power transmission equal to 0.5 SIL, 1 SIL, and 2 SIL.

The following line constants are assumed for the 765kV line:

R = 0.01 ohm/km
L = 8.35 × 10^{-4} H/km
G = 0
C = 12.78 × 10^{-9} F/km
Conductor Type: ACSR
Diameter: 35.1 mm
Rdc @ 20°C: 0.04 ohm/km

Number of sub-conductors / phase: 4
Number of circuit: 1
Frequency: 60 Hz
Latitude: 54°N
Longitude: 77°W
Time of day: 2 pm
Day: Dec 12
Meteorological Conditions:
Ambient temperature: 0°C
Wind speed: 1 m/s
Wind direction: 90° with respect to conductor
Sky condition: Clear sky

Solution

1. Surge Impedance Loading (SIL)
   Propagation constant is calculated,

$$\bar{\gamma} = \sqrt{\bar{Z} \cdot \bar{Y}}$$

$$\bar{\gamma} = \sqrt{(R + jL \cdot \omega)(G + jC \cdot \omega)}$$

$$\bar{\gamma} = 2 \cdot 10^{-5} + j1.2 \cdot 10^{-3}$$

The characteristic impedance is calculated,

$$\bar{Z}c = \sqrt{\frac{(R + jL \cdot \omega)}{(G + jC \cdot \omega)}}$$

$$Zc = 255.6 - j4.15$$

$$SIL = \frac{(765 \cdot 10^3)^2}{255.6} \cdot 10^{-6}$$

$$SIL = 2289 \text{ MVA}$$

2. Line ampacity is calculated by the program by following the procedure described in Chapter 3 for the specified transmission line conductor, meteorological conditions, and by consideration of four subconductors per transmission line phase.
   Ampacity/sub-conductor = 1920 A
   Line Ampacity = 4 × 1920 = 7680 A

Maximum power transmission capacity of the line = $\sqrt{3} \cdot 765 \cdot 7680 \cdot 10^{-3}$ = 5875.2 MVA
= 5875.2 MVA

To give an idea, this power is sufficient for many metropolitan cities.

3. Transmission line current as a function of line distance
The receiving end currentm I_r, is,

$$I_r = \frac{2289}{\sqrt{3} \cdot 765 \cdot 10^3}$$

Ir = 3455

Line to ground voltage V_r,

$$Vr = \frac{765 \cdot 10^3}{\sqrt{3}}$$

Vr = 441 kV

The transmission line current as a function of distance is obtained from Equation 10.28,

$$\bar{I}_x = \frac{\sinh\{(2.10^{-5} + j1.2 \cdot 10^{-3}) \cdot x\} \cdot 441 \cdot 10^3}{255.6 - j4.15} + \cosh\{(2.10^{-5} + j1.2 \cdot 10^{-3}) \cdot x\} \cdot 3455$$

The value of line current as a function of the distance from receiving end is shown in Figure 10.12. It is interesting to observe the variation of line current as a function of the distance. From this figure we can see that the thermal limit of the line is not exceeded even at two times the surge impedance loading (2 SIL) of the line.

4. Transmission line voltage as a function of line distance is calculated similarly from Equation 10.27 and is shown in Figure 10.13. It is interesting to observe the variation of line voltage as a function of line distance. From this figure we can see that there is substantial increase in line voltage at midpoint when power transfer is increased beyond the surge impedance load of the line.

10.6 PROTECTION

The effect of variable transmission line ratings on system protection requires careful evaluation to ensure proper functioning of the protective relaying system for both transmission and distribution lines. Transmission and distribution systems are generally provided with overcurrent and earth fault relays, impedance relays, differential relays, and voltage and underfrequency relays. These protective devices are designed to offer

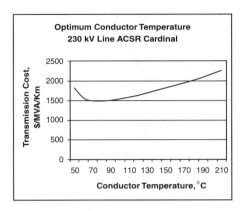

FIGURE 10.8 Optimum conductor temperature is 80°C as seen in the above figure.

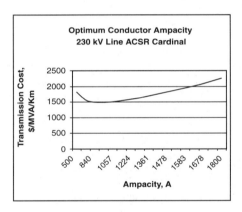

FIGURE 10.9 Optimum value of transmission line ampacity is 848 A.

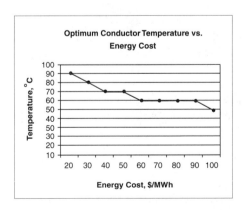

FIGURE 10.10 Optimum conductor temperature as a function of energy cost. As seen in this figure the optimum value of transmission line conductor temperature increases with lower electric energy cost.

Applications

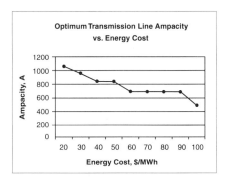

FIGURE 10.11 The optimum value of transmission line ampacity increases with lower electric energy cost.

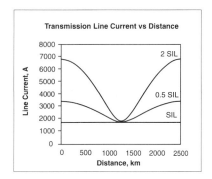

FIGURE 10.12 Variation of transmission line current as a function of distance for different power transmission levels.

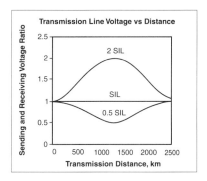

FIGURE 10.13 Variation of transmission line voltage as a function of distance for different power transmission levels.

protection to transmission lines, distribution feeders, transformers, substation bus-bars, and generators, as well as the loads they serve. For satisfactory functioning of the protection system, the system must be able to distinguish between permissible overload current and a fault current to avoid faulty tripping during an acceptable overload condition.

To ensure fault discrimination, protection relays are time-coordinated so that the circuit breaker closest to the fault opens first. Backup protection is provided so that if the breaker closest to the fault fails to operate, the next breaker will open. Automatic reclosures are also provided on most circuits for automatic recovery from temporary faults. Reclosurers are circuit breakers that close automatically at predetermined intervals after opening a circuit to eliminate faults that are temporary in nature. For all of the above protection schemes, the magnitude of the overcurrent and relay operation times are determined from network short-circuit studies with proper prefault and postfault line operating conditions.

Traditionally, the relay pickup current setting for overload protection was determined by static line ratings. In a static system, the ratings of transmission lines, transformers, and other substation current-carrying equipment are usually considered to be constant during a season, resulting in winter and summer ratings. For networks having dynamic line ratings, the protective relaying settings will have to be updated continuously, preferably on a real-time basis. This is the subject of adaptive relaying and beyond the scope of this book. The interested reader is referred to an excellent book on the subject of adaptive computer relaying (Phadke et al.1988) and other excellent technical papers in IEEE, Cigré, and similar conferences.

10.7 CHAPTER SUMMARY

Applications of powerline ampacity system to power system economic operation, load-flow, generator stability, transmission line planning, and design considerations of overhead powerlines in view of powerline ampacity were presented in this chapter. A formulation of the optimal power flow problem was given to show the significance of transmission line dynamic thermal ratings in the economic operation of an interconnected electric power system having diverse generation sources. Electricity production costs were evaluated with static and dynamic line rating in a transmission network having diverse generation sources. Results were presented to show the savings in electricity production cost achieved by the dynamic rating of transmission lines using LINEAMPS.

Examples were provided in the chapter to show the impact of high transmission line ampacity on steady-state generator stability, and dynamic and transient stability. It was shown that power system stability limits are enhanced by dynamic line ratings.

The application of a powerline ampacity system in the planning and design of new overhead lines includes the selection of optimum conductor size, optimum current density, and the evaluation of alternative conductor designs. For this purpose, a complete formulation of transmission line economics was presented. Economic analysis of existing lines show that the construction of new lines, or the reconductoring of existing wires, may be postponed in many cases by dynamic line ratings, with substantial cost savings.

Appendix 10
Transmission Line Equations

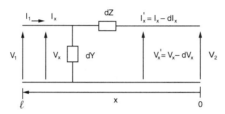

FIGURE A10.1 Long transmission line model.

Applying Kirchoff's law to a small section dx (dx << λ) of the line we have,

$$d\overline{V}_x = \overline{V}_x - \overline{V}'_x = d\overline{Z}(\overline{I}_x - d\overline{I}_x) \approx d\overline{Z} \cdot \overline{I}_x = \overline{Z} \cdot \overline{I}_x dx \tag{A10.1}$$

$$d\overline{I}_x = \overline{I}_x - \overline{I}'_x = d\overline{Y} \cdot \overline{V}_x = \overline{Y} \cdot \overline{V}_x dx \tag{A10.2}$$

where the product of differential quantities are neglected. From above, the following second-order linear differential equations are obtained,

Voltage,

$$\frac{d^2 V_x}{dx} = \overline{Z} \cdot \overline{Y} \cdot \overline{V}_x = \overline{\gamma}^2 \overline{V}_x \tag{A10.3}$$

$$\frac{d^2 V_x}{dx} - \overline{\gamma}^2 \overline{V}_x = 0 \tag{A10.4}$$

Current,

$$\frac{d^2 \overline{I}_x}{dx} = \overline{Z} \cdot \overline{Y} \cdot \overline{I}_x = \overline{\gamma}^2 \overline{I}_x \tag{A10.5}$$

$$\frac{d^2 \bar{I}_x}{dx} - \bar{\gamma}^2 \bar{I}_x = 0 \qquad (A10.6)$$

where the propagation constant $\bar{\gamma}$ is given by,

$$\bar{\gamma} = \sqrt{\bar{Z} \cdot \bar{Y}} \qquad (A10.7)$$

The solutions of differential equations (10.4), (10.6) are,

$$\bar{V}_x = k_i e^{\bar{\gamma}x} + k_r e^{-\bar{\gamma}x} \qquad (A10.8)$$

$$\bar{I}_x = \frac{k_i}{\bar{Z}c} e^{\bar{\gamma}x} - \frac{k_r}{\bar{Z}c} e^{-\bar{\gamma}x} \qquad (A10.9)$$

at x = 0 we have,

$$\bar{V}_2 = \bar{k}_i + \bar{k}_r \qquad (A10.10)$$

$$\bar{Z}_c \bar{I}_2 = \bar{k}_i - \bar{k}_r \qquad (A10.11)$$

$$\bar{k}_i = \frac{\bar{V}_2 + \bar{Z}_c \bar{I}_2}{2} \qquad (A10.12)$$

$$\bar{k}_r = \frac{\bar{V}_2 - \bar{Z}_c \bar{I}_2}{2} \qquad (A10.13)$$

$$\bar{V}_x = \frac{\bar{V}_2 \left(e^{\bar{\gamma}x} + e^{-\bar{\gamma}x} \right)}{2} + \bar{Z}_c \bar{I}_2 \left(e^{\bar{\gamma}x} - e^{-\bar{\gamma}x} \right) \qquad (A10.14)$$

$$\bar{I}_x = \frac{\bar{V}_2 \left(e^{\bar{\gamma}x} + e^{-\bar{\gamma}x} \right)}{2 \bar{Z}c} + \frac{\bar{I}_2}{2} \left(e^{\bar{\gamma}x} - e^{-\bar{\gamma}x} \right) \qquad (A10.15)$$

$$\bar{V}_x = \cosh(\bar{\gamma}x) \bar{V}_2 + \bar{Z}c \cdot \sinh(\bar{\gamma}x) \bar{I}_2 \qquad (A10.16)$$

$$\bar{I}_x = \frac{\sinh(\bar{\gamma}x) \bar{V}_2}{\bar{Z}c} + \cosh(\bar{\gamma}x) \bar{I}_2 \qquad (A10.17)$$

substituting x = ℓ, we obtain the sending end voltage current as follows,

Appendix 10 Transmission Line Equations

$$\overline{V}_s = \cosh(\overline{\gamma} \cdot \ell)\overline{V}_2 + \overline{Z}c \cdot \sinh(\overline{\gamma} \cdot \ell)\overline{I}_2 \qquad (A10.18)$$

$$\overline{I}_s = \frac{\sinh(\overline{\gamma} \cdot \ell)\overline{V}_2}{\overline{Z}c} + \cosh(\overline{\gamma} \cdot \ell)\overline{I}_2 \qquad (A10.19)$$

Equations (A10.18) and (A10.19) in matrix form relating sending end voltage, Vs, and current, Is, to receiving end voltage, Vr, and current, Ir, is,

$$\begin{bmatrix} \overline{V}s \\ \overline{I}s \end{bmatrix} = \begin{bmatrix} \overline{A} & \overline{B} \\ \overline{C} & \overline{D} \end{bmatrix} \begin{bmatrix} \overline{V}r \\ \overline{I}r \end{bmatrix} \qquad (A10.20)$$

where,

$$\overline{A} = \cosh(\overline{\gamma} \cdot \ell)$$

$$\overline{B} = \sinh(\overline{\gamma} \cdot \ell)\overline{I}_r$$

$$\overline{C} = \frac{\sinh(\overline{\gamma} \cdot \ell)}{\overline{Z}c}$$

$$\overline{D} = \cosh(\overline{\gamma} \cdot \ell)$$

11 Summary, Future Plans and Conclusion

11.1 SUMMARY

As the demand for electricity grows in all regions of the world, there is greater need to develop better ways of operating existing Transmission and Distribution (T&D) networks for more efficient utilization of existing facilities. Historically, when demand increased, electric utilities added new T&D capacity by the construction of new lines and substations. Due to public concern about protecting the environment, and to population growth, public authorities are paying greater attention to properly locating electric power facilities and proper land use. As a result, there is greater R&D effort by power companies and concerned government agencies to minimize environmental impact, increase energy efficiency, and improve land use.

In this book, a complete system of rating overhead powerlines is developed giving theory, algorithms, and a methodology suitable for implementation in a computer program. The program is not only an operational tool enabling better utilization of existing transmission and distribution facilities, but is also a valuable planning tool for line maintenance, and a design tool for the construction of future T&D facilities. The line ampacity system described in this book offers an integrated line ampacity system comprising a transmission line model, a conductor model, and a weather model for the first time. By developing a system of rating overhead lines as a function of forecasting weather conditions by an object-model and expert system, a very user-friendly program is realized that is easily implemented in all geographic regions.

Chapter 1 introduced the subject of transmission line ampacity and presented the line ampacity problem. It is stated that voltage and stability limits can be improved by control of reactive power and/or boosting voltage levels by transformer action. Therefore, in many cases the transport capacity of overhead powerlines is limited only by the thermal rating of powerline conductors.

Chapter 2 described line rating methods from the early works of Ampere, Faraday, and several researchers from different countries who were concerned with the problem of transmission line ampacity, and who offered solutions to increase line capacity. Utility line rating practices, and online and offline methods are discussed, including real-time ratings, forecast ratings, probabilistic ratings, and static line ratings.

Chapter 3 provided a complete theory of conductor thermal ratings. First, a three-dimensional conductor thermal model is developed, and then steady-state, dynamic, and transient thermal rating models are developed from it. The concept of steady, dynamic, and transient ratings are introduced for the first time in this chapter and are related to machine stability in Chapter 10.

Chapter 4 described experimental research in the laboratory, on outdoor transmission line test span, as well as field measurements on real transmission line circuits. This chapter presents up-to-date knowledge in the field of line ampacity calculations by comparing different line ampacity calculation methods. Theoretical results that were obtained by calculation from the transmission line conductor thermal models developed in Chapter 3 are compared to IEEE and Cigré standards with excellent agreement.

Chapter 5 presented elevated temperature effects of different types of transmission line conductors. Experimentally-determined empirical models of loss of tensile strength of conductors and permanent elongation due to elevated temperature operations were presented with results from each model. A probability method of calculation of transmission line sag and tension was presented for the first time. The probability distribution of conductor temperature in service was obtained by a Monte-Carlo simulation of time series stochastic models. Computer algorithms were presented for the recursive estimation of loss of strength and permanent elongation. Results were compared with data from industry standards and transmission line field data with excellent agreement.

Chapter 6 presented a study of the electric and magnetic fields from high-voltage power transmission lines. This aspect of transmission line ampacity is significant because there is little previous work carried out in this direction. A complete theory with examples of the method of calculation of the electric and magnetic fields from a line were presented to form a clear understanding of the subject. Even though there is no evidence of any significant environmental impact by EMF due to increased transmission line current, measures are suggested for the reduction of electric and magnetic fields from transmission lines by new conductor configurations, and by active and passive shielding.

Chapter 7 described various approaches to weather modeling for the prediction of line ampacity. Since weather is an important parameter in transmission line ampacity calculations, the development of weather models of ambient temperature, wind speed, wind direction, and solar radiation were presented. First, statistical weather models based upon time-series analysis of National Weather Service forecasts were developed. Then, neural network modeling was presented for forecasting and pattern recognition. Examples of hourly values of future meteorological conditions generated from the models are given. Application of fuzzy set modeling of transmission line ampacity is also given.

Chapter 8 described the computer modeling method of the line ampacity system and its implementation in a computer program called LINEAMPS. A significant contribution made by the program is that it not only predicts line ampacity, but also offers a complete environment for the management, planning, and operation of overhead powerlines. The program offers a user-configurable transmission line database for the first time, thanks to recent developments in object oriented modeling technology (Cox 2000). The database is comprised of powerlines, different types of powerline conductors, and weather stations for any geographic region.

Summary, Future Plans and Conclusion

LINEAMPS provides the ability to create new line, new conductors and new weather stations. It is a user-friendly program created by the application of artificial intelligence using object-model and expert system rules. Users of the program have less chance of making errors in data input because expert rules check user input and explain error messages. It is also a contribution to the field of cognitive science applied to electric power transmission where a computer program resembles a human expert.

Real examples* of computer modeling of the line ampacity system in two different geographic regions of the world are presented in Chapter 8. It shows the suitability of the program in all geographic regions of the world. As of this writing, LINEAMPS is being used or evaluated in the following power companies in the different regions of the world.

- TransPower, New Zealand (Figure 11.5)
- Korea Electric Power Company, South Korea (Figure 11.4)
- Electricité de France (EDF), Paris, France (Figure 11.3)
- Hydro Quebec, Canada (Figure 11.2)

Chapter 9 discussed the state of the art in the development of power semiconductor devices for FACTS applications. The various types of FACTS devices are described in this chapter, including Static Var Compensating (SVC) devices, Thyristor Controlled Series Compensation (TCSC), STATCOM, UPFC, and Superconducting Magnetic Energy Storage (SMES). An example of an actual FACTS installation in a high-voltage substation is also provided.

Chapter 10 presented applications of the line ampacity system to transmission system planning, economic power system operation, and load-flow and generator stability issues. Cost saving by the implementation of a variable line rating system in an interconnected electric network was presented by example. An important feature of the new line ampacity system in a competitive electricity supply market is the ability to forecast hourly values of line ampacity up to seven days in advance, enabling advance purchase and sale of electricity.

A formulation of the optimal power flow problem was presented in Chapter 10 to show the significance of transmission line dynamic thermal ratings to the economic operation of an electric power system with diverse generation sources. The application of the powerline ampacity system in the planning and design of new overhead lines includes the selection of optimum conductor size, optimum current density, and the evaluation of alternative conductor designs. For this purpose, a complete formulation of transmission line economics was presented. Economic analysis of existing lines shows that the construction of new lines or the reconductoring of existing wires can be postponed in many cases by dynamic line ratings, with substantial cost savings. The cost of electricity may be reduced by greater utilization of economy energy sources and, last but not least, there is a lowering of environmental impact, both visual and ecological.

---

* *LINEAMPS for New Zealand User Manual*, 1995, describes modeling transmission lines and weather stations for the region of New Zealand, North and South Island.
LINEAMPS for Korea User Manual, 1996, describes modeling transmission lines and weather stations for the region of South Korea.

11.2 MAIN CONTRIBUTIONS

The Concept of Steady, Dynamic and Transient Line Rating

Most of the literature on transmission line conductor thermal rating considers the short-term thermal rating of a conductor as its transient thermal rating. In effect, it would be more appropriate to call short-term rating "dynamic line rating," because the time span is of the order of several seconds, which corresponds well with the time span considered in dynamic stability studies on generators.

The transient thermal rating of the conductor is considered when there is a short-circuit or lightning currents. Therefore, the time span of transients is much shorter, generally in the order of milliseconds, and corresponds well with the time span considered in transient stability studies on generators. Thus, the concept of steady-state rating, dynamic rating,* and transient line rating were introduced for the first time in relation to generation stability as presented in Chapter 10.

The above terminology of the different transmission line ratings is consistent with other areas of the electric power system relating to the steady-state stability, dynamic stability, and transient stability of electrical generators. The steady-state transmission line conductor thermal rating is applicable to the condition of steady-state operation of a power system; dynamic thermal rating is applicable during power system dynamic operations; and transient thermal rating is applicable during power system transient operating conditions.

Three-Dimensional Conductor Thermal Model

A three-dimensional conductor thermal model was used in a previous wind tunnel study conducted by the author at PG&E to determine radial temperature distribution inside a transmission line conductor (Hall, Savoullis, Deb, 1988). A three-dimensional conductor thermal model was developed in a recent report (Cigré, 1997) for the calculation of transmission line ampacity.

In Chapter 4, the differential equation of conductor temperature was developed from a three-dimensional conductor thermal model. The main reason for selecting this model is that it enables the calculation of radial temperature distribution and the average conductor temperature within the conductor. As we may recall from Chapter 5, the sag and tension of a transmission line conductor is calculated from the average conductor temperature.

Transmission Line Risk Evaluation

By evaluating weather conditions and transmission line temperatures in the region of Detroit (Davis, 1977), it was shown that transmission line capacity was greater than static line capacity for a large percentage of the time in that region. Davis also showed that static line rating is not risk free, and there exists a small percentage of time when maximum conductor temperature is exceeded even by static line rating.

---

* For a discussion on dynamic rating terminology, please see the discussion contribution in the references (Hall, Deb, 1988a).

Similarly, the research carried out at PG&E in California also showed that actual transmission line ampacity is significantly greater than static line ratings (Hall and Deb, 1988b) for a large percentage of time. A transmission line risk analysis study was carried out at PG&E which showed there is minimum risk to a dynamic line rating system (PG&E Report. 1989).* The results presented in this book show that there is minimum risk as conductor sag; loss of strength and EMF were calculated and found to be within acceptable limits.

Real-Time Rating

Real-time line rating systems (Davis, 1977), (Seppa, T.O., 1998), (Soto et al., 1998), (Deb, 1998) calculate line ampacity by the measurement of weather variables, conductor temperature, and/or conductor tension. In Davis's system, real-time line ratings were calculated by monitoring conductor temperature with sensors installed at different sections of the line. In the PG&E system (Mauldin et al., 1988), (Steeley et al., 1991), (Cibulka et al., 1992), line ampacity is calculated by monitoring ambient temperature only. The PG&E line rating system also offered one to 24 hours ahead forecast rating capability. For this reason, real-time stochastic and deterministic models were developed by the author at PG&E to forecast transmission line ampacity up to 24 hours in advance (Hall, Deb, 1988b), (Steeley et al., 1991).

Another real-time line rating system calculates ampacity by the measurement of conductor tension (Seppa, T.O., 1998). The relationship between transmission line conductor temperature and tension is presented in the conductor change of state equation in Chapter 5. By the application of the change of state equation we can easily calculate conductor temperature if conductor tension is measured. Ampacity is then calculated from conductor temperature by the application of the conductor thermal models presented in the Chapter 3. It is difficult to obtain forecast ratings by monitoring conductor tension or conductor temperature alone. It is easier to forecast line ratings by forecasting weather conditions.

Installation of temperature sensors or tension monitors requires taking the line out of service for periodic maintenance. It also requires communication with sensors installed on the line to the utility power control center computer where line ampacity is calculated by a program. The prediction of line ampacity by monitoring weather conditions does not require any new hardware to be installed on a transmission line. Weather conditions are generally monitored in electric utility power systems for other reasons, such as load forecasting, etc. Therefore, real-time weather data is available in most electric power company control centers at no additional cost. It is also expensive to install temperature or tension monitoring devices on all transmission lines. Thus, at the present time, these devices are installed on a limited number of heavily-loaded critical transmission circuits.**

For the above reasons, a line monitoring system is expected to complement the more general-purpose line rating system developed in this book. The line rating

---

* The report presents a quantitative analysis of the risk of an ambient adjusted line rating system developed at PG&E.
** (Waldorf, Stephen, P., Engelhardt, John S., 1998). This paper describes real-time ratings of critical transmission circuits by monitoring temperature of overhead lines, switchgear, and power transformers.

methodology presented in this book offers a more economical system that can be easily implemented for the rating of all lines in all geographic regions.* (See Figures 11.2–11.5). The Spanish system (Soto et al., 1998) is already beginning to follow this approach. However, their system lacks forecast rating capability, weather modeling, and the ability to make new conductors, new weather stations, and new transmission lines, which are the objects of the author's patent (Deb. 1999).**

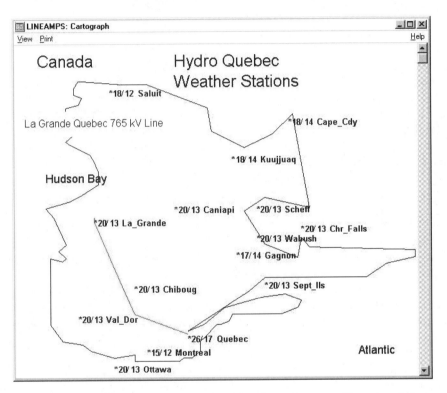

FIGURE 11.2 LINEAMPS for Hydro Quebec. Map shows the geographic region of Quebec, Canada and weather stations created by program. A 765 kV transmission line from La Grande hydroelectric stations to Montreal, Quebec is shown in the figure.

---

* *LINEAMPS for Hydro-Quebec, Canada, Software Users Guide*, 1998, describes the line campacity system developed for Hydro-Quebec in Canada.
LINEAMPS for EDF, France, Software Users Guide, 1998, describes the line ampacity system developed for EDF, France.
LINEAMPS for TransPower, New Zealand, Software Users Guide, 1996, describes the line ampacity system developed for TransPower, NZ.
LINEAMPS for KEPCO, S. Korea, Software Users Guide, 1996, describes the line ampacity system developed for TransPower, NZ.
** Anjan K. Deb, 1999. "Object-oriented line ampacity expert system." U.S. Patent.

Forecast Rating

In the line ampacity system developed in this book, ampacity is forecast up to seven days in advance by adjusting the LINEAMPS weather model to National Weather Service forecasts. Hourly values of ambient temperature, wind speed, and solar radiation are generated by AmbientGen, WindGen, and SolarGen methods in each weather station object as described in Chapter 7. At the present time, line ampacity forecasts are limited to seven days in advance because weather forecasts are less accurate beyond that period.*

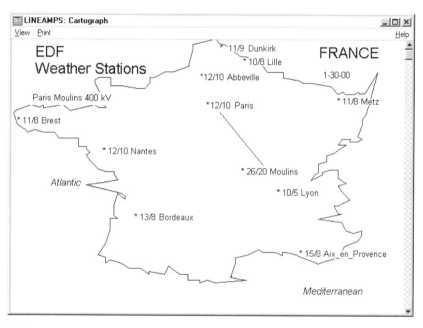

FIGURE 11.3 LINEAMPS for EDF, France. EDF is the national electric company of France. Weather stations are created by the LINEAMPS program. A 400 kV transmission line from Paris to Bordeaux created by program is also shown in the figure.

Weather and Line Current Modeling

A time-series stochastic model of ambient temperature for the prediction of transmission line conductor temperatures was first presented by the author (Deb, 1985) at the (Cigré Symposium 1985)** on High Currents and (Hall, Deb, 1988b). The Box-Jenkins forecasting model is quite accurate, but the model coefficients could not be easily adapted in real-time. Therefore, a recursive least square estimation model was developed at PG&E suitable for real-time calculation (Steeley, Norris, and Deb, 1991), (Cibulka, Steeley, Deb, 1992). Statistical models based on hourly differences of ambient temperature were also proposed by others (Douglass, 1986), (Foss, Maraio,

---

* Weather section, *USA Today*, November 23, 1998.
** The Cigré Symposium was exclusively devoted to the subject of transmission line ampacity.

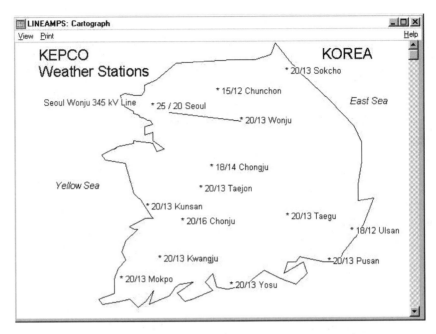

FIGURE 11.4 LINEAMPS for KEPCO, South Korea. KEPCO is the national electric company of South Korea. Weather stations are created by the LINEAMPS program. A 345 kV transmission line from Seoul to Wonju, also created by program, is shown in the figure.

1989) but it was shown in a discussion contribution* that real-time recursive estimation was more accurate. The following forecasting models were evaluated:

- Fourier series model
- Recursive least square estimation algorithm
- Kalman Filter
- Neural network

The results presented in the Chapter 3 and Chapter 4 show the accuracy of each model. A transmission line current model is useful for the prediction of conductor temperature, and may be developed by using the forecasting models mentioned above (Deb et al., 1985).

The Fourier series model of ambient temperature and wind speed is used in the LINEAMPS program. This model is recommended for its simplicity, and is suitable when a general-purpose weather forecast of a region is available.** A neural network is useful for weather pattern recognition, as shown in Section 4.3. A recursive

---

* Discussion contribution by Anjan K. Deb and J.F. Hall to the IEEE paper (Douglass, D.A., 1988).
** General-purpose weather forecast issued by the National Weather Service includes daily maximum and minimum values of ambient temperature for three to seven days in advance. The data is generally published in daily newspapers as well as in several Internet web sites: www.intellicast.com, www.nws.noaa.gov.

Summary, Future Plans and Conclusion

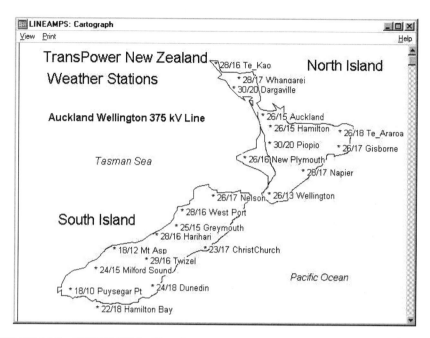

FIGURE 11.5 LINEAMPS for TransPower, New Zealand. TransPower operates the transmission grid in New Zealand. The weather stations shown in the above map are created by the LINEAMPS program. A 375 kV transmission line from Auckland to Wellington in the North Island of New Zealand is also shown in the figure.

estimation algorithm and Kalman filter are suitable for real-time forecasting of transmission line ampacity on an hourly basis.

Transmission Line Object-Model

Object-oriented modeling is a new way of developing computer programs that makes use of inheritance, polymorphism, communication by message, and delegation of messages (Booch, G., 1993), (Cox, E., 2000), (Kappa-PC).* The object-model is particularly suitable for electric power system applications for the modeling of transmission lines, generators, and other substation equipment, and is seriously considered by electric power companies (*MPS Review* Article, 1998). The object-model approach is used to model transmission lines, weather stations and transmission line conductors in the LINEAMPS program (Deb, 1995, 1997, 1998). By using the object-model approach, it is shown in Chapter 5 how transmission line objects are easily created with attributes and behavior by class inheritance.

Examples of the object-model that were created for the regions of New Zealand and South Korea are presented in Chapter 5. It is seen in this chapter how a systematic method of transmission line ampacity system data classification is developed. Transmission lines are classified by voltage levels, weather station data

---

* Kappa-PC ver 2.4 provides connection to industry standard databases by ODBC, which enables the LINEAMPS program to access transmission line and weather data.

Line Ampacity Expert-System

The declarative style of programming by rules was used to develop the LINEAMPS program. There are several advantages of programming by rules instead of purely by procedures (Kronfeld, Kevin M., Tribble, Alan C., 1998). For example, expert systems use an inference engine to make decisions, while in a procedural language the programmer writes the code for decision making. The LINEAMPS program uses an expert system to check user input data. Similarly, there are other expert systems developed for the power industry for fault diagnosis (Taylor et al., 1998),* intelligent tutoring systems (Negnevitsky, 1998), and power quality (Kennedy, B. 2000).

LINEAMPS is the first transmission line expert system using object-oriented modeling and rules (Deb, 1995). As shown in Chapter 8, rules of thumb and practical knowledge are easily implemented by an expert system. Examples are presented to show how the program calculates powerline ampacity by objects and rules, checks user input data, and explains error messages like a true expert.

Experimental Verification of Transmission Line Ampacity

Experimental work carried out at the PG&E wind tunnel to verify the conductor thermal model (Hall, Savoullis, Deb, 1988) is described in Chapter 6. A new equation was developed from wind tunnel data to determine the Nusselt number as a function of the Reynolds number (Ozisik, M.N., 1985) for the calculation of convection cooling by wind. Results of steady state ampacity and dynamic and transient ampacity obtained from the LINEAMPS program are presented in a table and compared to wind tunnel data and other data compiled from various power companies (Urbain J.P., 1998), (PG&E Standard, 1978)** and a conductor manufacturer's catalog (Southwire, 1994)*** with excellent agreement. The values of steady-state ampacity and dynamic ampacity calculated by the LINEAMPS program also compared well with the IEEE standard and a recent Cigré report.

Hourly values of an actual 345kV transmission line ampacity was available from Commonwealth Edison Company**** (ComEd), Chicago. The line ampacity data was obtained from a real-time line rating system operated by ComEd. A comparison of this data from 8:00 a.m. to 4:00 p.m. showed that LINEAMPS ratings never exceeded real-time ratings at any time. LINEAMPS also accurately predicted the lowest rating at 12:00 noon.

---

* (Taylor et al., 1998) provides a list of expert systems developed for the power industry worldwide.
** (PG&E Standard, 1978) gives the static line rating for summer and winter of all conductors used by PG&E for normal and emergency conditions, and the corresponding assumptions of meteorological conditions.
*** (Southwire, 1994) provides the ampacity of all commonly used conductors in the U.S., and the corresponding assumptions of meteorological conditions.
**** Thanks are due to Mr. S. Nandi, Technical Expert, Transmission Line Thermal Rating Studies, Commonwealth Edison, Chicago, IL.

Summary, Future Plans and Conclusion

The above example showed that LINEAMPS ratings have minimum risk of exceeding allowable maximum conductor temperatures.

Effects of Higher Transmission Line Ampacity

When transmission line ampacity is increased, it is necessary to properly evaluate the thermal effects of the powerline conductor as well as the electric and magnetic fields of the transmission line. A unified approach to the modeling and evaluation of the effects of higher transmission line ampacity is presented in Chapter 5. The thermal effects include conductor loss of strength and permanent elongation of the transmission line conductor. These effects are evaluated recursively from the probability distribution of conductor temperature by using Morgan's equation (Morgan, 1978).

Morgan calculated the loss of tensile strength of conductor by a percentile method. The same method was recently used to calculate the thermal deterioration of powerline conductors in service in Japan (Mizuno et al., 1998, 2000). In their study, the reduction in tensile strength of the conductor was used as the index of thermal deterioration of the conductor. They showed by simulation that conductor loss of strength is less than 1% by static line rating. They also showed that the frequency distribution of a conductor should consider the relationship between weather and line current. In Chapter 5, the probability distribution of conductor temperature is generated by a Monte-Carlo simulation of weather data from time-series stochastic models and transmission line currents which consider the correlation between weather and load current.

In addition to the loss of tensile strength, conductor sag and tension are also affected by elevated temperature operation due to high currents (Cigré, 1978). A new method is developed to determine the sag and tension of overhead line conductors with elevated temperature effects. The accuracy of sag-tension program was tested with ALCOA program (Lankford, 1989), STESS program (CEA Report, 1980), and KEPCO's transmission line field data (Wook, Choi, Deb, 1997). Results are presented which compare well with all of the above data.

The magnetic field at ground level of a transmission line increases with line ampacity and conductor sag (Rashkes, Lordan, 1998). Most studies on magnetic fields from powerlines are conducted with typical transmission line currents equal to or less than static line ratings. The magnetic fields of conductors with dynamic ampacity are presented in Chapter 6.

Typical powerline configurations are evaluated to show the magnetic fields with high currents. It is shown that the magnetic fields of overhead transmission lines with dynamic line ampacity are within acceptable limits. Line design and EMF mitigation methods are suggested to lower transmission line magnetic fields in sensitive areas (Böhme et al., 1998). The electric field of a high-voltage line at ground level does not depend upon transmission line ampacity if the maximum design temperature of the conductor is not exceeded and minimum conductor-to-ground distance is maintained.

Line Ampacity Applications

The application of a dynamic line rating system in the economic operation of an electric power system was first presented by (Hall, J. F., Deb, Anjan K., 1988a), and later on by (Deb, 1994). Deb developed the economic analysis method of dynamic line rating system with adaptive forecasting to demonstrate power system operational cost savings by dynamic thermal rating. Substantial capital cost savings by the deferment of capital investment required for the construction of new lines and environmental benefits were shown, as fewer lines are required.

A recent study (Yalcinov, T. and Short, M. J., 1998) on generation system cost optimization by neural network now confirms that substantial cost saving is possible by increasing transmission line capacity. The theory and mathematical model of a hypothetical utility system are presented in this book to demonstrate electric power system economy achieved by the new line ampacity system. In addition to the solution of the optimal power flow problem by a classical solution of the nonlinear optimization problem (Bergen, A., 1986), artificial neural network (Lee, K.Y. et al., 1998), and genetic algorithm GA (Wong, Yuryevich, 1998) were developed in the industry to obtain faster and more efficient solutions to the optimal power flow problem.

11.3 SUGGESTIONS FOR FUTURE WORK

Overhead powerlines constitute the single most important component of the electrical power system. Therefore, the contributions made in this book to increase line capacity of existing overhead power transmission and distribution lines will have a significant impact in the improvement of the overall performance of the electrical power system. The power system will continue to evolve as our demand for electricity increases in a fair and competitive environment based on free market principles.

Power plants and transmission lines require substantial investments, which must be carefully evaluated before new facilities are added. Following is a list of areas that require further research for optimum utilization of existing assets, and the development of new transmission line technologies with greater emphasis on renewable energy sources and minimum environmental impact. A plan to develop the line ampacity system further for application in a deregulated electricity production, supply, and distribution environment is also proposed. It is hoped that these studies will lead to the optimum utilization of all resources.

Overcoming Limitations of Existing AC Networks by the Improvement of Power System Stability, Economy, Energy Transfer, and Reliability

In the past, the ampacity of existing overhead powerlines could not be fully utilized because existing Alternating Current (AC) circuits are mostly composed of passive elements having very little controllability. Therefore, when demand increases, existing network control methods are not sufficient to properly accommodate increased power flows. As a result, certain lines are more heavily loaded and stability margins are reduced. Due to the difficulty of controlling power flow by existing methods,

Summary, Future Plans and Conclusion

FACTS devices are being developed and are used to control existing T&D networks. These devices are presently installed at a number of locations worldwide, at all voltage levels up to 800 kV, with varying capacities.

FACTS Technology

Due to the development of FACTS technology it is now possible to fully exploit existing transmission line capacities. By locating FACTS devices at suitable locations on existing networks, it is now possible to exercise a range of control over the AC network in a manner that was not possible before. Thyristor controlled shunt and series capacitors are now widely used in the electric power system to increase the transmission capacity of existing lines. Similarly, SMES devices are installed for faster response and voltage support, and to maintain stability.

Increasing transmission line capacity by dynamic thermal rating adjusted to actual weather conditions is not only useful for economic energy transfer, it is also beneficial for the enhancement of power system reliability, including dynamic and transient stability. In addition to the SVC system, there are other new developments such as UPFC, PowerFormer, and Superconducting Magnetic Energy Storage that will enable even greater control over power flow through transmission lines.

The above examples demonstrate the importance of developing a dynamic transmission line thermal rating system with FACTS for the improvement of power system stability, and increasing the power transfer capability of existing lines to achieve greater economy.

Greater Utilization of Renewable Energy Sources

It has been said that the electric power system is the most complex system ever created by human beings. Since transmission lines constitute the most important component of a power system, the powerline ampacity system presented in this book will continue to develop further and is expected to play a vital role in the planning and operation of existing and future powerlines in all regions.

With public concern about global warming, there will be even tighter control on the emission levels of electricity generated by burning fossil fuels. There is now greater emphasis on producing electricity from renewable energy sources including solar, wind, and hydroelectricity. Therefore, an interconnected transmission network is essential to connect all of these sources of energy.

The world has an abundance of renewable energy sources. For example, hydroelectric reserves in Alaska and Canada in North America, and in the Amazon in South America, are yet to be developed. Similarly, there exists a large potential for solar energy development in the Sahara desert. When renewable energy sources are fully developed, it is expected to meet all of our energy requirements for this planet. As these sources of energy are generally at remote locations, Extra High Voltage (EHV) transmission lines will be required to bring electricity from remote locations to major metropolitan areas and industrial centers.

In the past, electricity could be transmitted efficiently to a distance of about 2500 km. Research from the International Conference on Large High Voltage

Networks (Cigré) now shows that it is technically and economically feasible to transport electricity to about 7000 km by Ultra-High Voltage (UHV) lines (*IEEE Power Engineering Review*, 1998).* This modern trend in electricity transmission is already beginning to happen. EHV lines are presently transporting low cost electricity to New York generated from renewable energy sources in Canada.** Similarly, EHV and HVDC lines are planned to bring low-cost hydroelectricity from Siberia to Japan.

Interconnection of a high-voltage power transmission network will also enable the flow of electricity across time zones, thereby taking advantage of the difference in time-of-day demand for electricity. As transmission lines extend beyond political boundaries, it is hoped this will lead to greater cooperation between nations, bringing greater peace and prosperity. With further development of the high-voltage transmission network to accommodate the diverse sources of energy, and interconnection with neighboring countries, the powerline ampacity system will become more useful for the precise control of the current flowing through transmission line wires.

Development of New Transmission Line Technology

New transmission line technology is required to achieve greater energy efficiency and reliability of supply by interconnection, with minimum environmental impact. Advanced systems of communication are being developed using powerline communication by fiber optics (J.P. Bonicel, O. Tatat, 1998), which is free from electromagnetic disturbances. The fiber optic core can be easily used as a continuous wire temperature sensor, thus eliminating the uncertainty of present-day line ratings. Furthermore, the Internet will be used extensively for utility data communication, and a utility transmission line network could become a part of the Internet for the reliable transmission of data.

Modern high-voltage transmission networks will be used to carry electric energy and data for the precise control of power flow through the wires, in addition to other business and commercial uses. Here, again, there is greater opportunity for the line ampacity system to develop further. It will send signals through the powerline communication network to remote power electronics (FACTS) devices for the precise control of power flow through the lines.

In the field of overhead transmission line conductor technology, more research is required to develop high temperature conductors using advanced aluminum alloys (Wook, Choi, Deb, 1997) for higher transmission capacity, and compact conductors to lower transmission losses (Jean-Luc Bousquet et al., 1997). Conductors with fiber-optic communication technology will be required for long-distance communication with minimum number of booster stations.

---

* *IEEE Power Engineering Review*, 1998, presents articles from several authors in different countries for the development of renewable energy sources and UHV transmission to meet the energy requirements of this planet in the 21st century.
** North American Electric Reliability (NERC) Transmission Map. NERC, 101 College Rd, Princeton, NJ 08540.

Summary, Future Plans and Conclusion 223

Environmental Impact

The world is paying greater attention to the environment (Alexandrov, G.N. et al., Russia, 1998), (Awad et al., Egypt, 1998)* by greater consideration of factors such as EMF from powerlines, land use and ecological effects, and the emission level of fossil fuel generators. In Chapter 6, the electric and magnetic fields of typical transmission line configurations is calculated by the application of Maxwell's equation. It is shown that the magnetic field of a line is minimum at ground level, and well below acceptable limits. EMF mitigation measures are also suggested by compact line design, modified phase configuration, and active shielding to minimize EMF near sensitive areas.

Underground Transmission

Underground lines are recommended for aesthetic reasons in cities and scenic areas. Since they are underground, they have no visual impact on the surrounding environment. Underground transmission lines have no electric fields on ground surfaces since the outer layer of the cable is generally connected to ground and remains at ground potential. Unless properly shielded, there is a magnetic field from the cable at ground surface.

There are certain disadvantages of underground cables.** Underground cables take up a much greater area of land than that needed for overhead lines of the same capacity. To place a 400,000-volt line underground would typically involve digging a trench the width of a three-lane road and $1^1/_2$ meters in depth to accommodate up to 12 separate cables. The space is needed because the high-voltage cable generates a great deal of heat — equivalent to a one-bar electric fire every two meters. Whereas the surrounding air directly cools overhead lines, underground cables have to be well spaced to allow for natural cooling and to avoid overheating. Additional land is needed at cable terminals where underground cables are joined to overhead lines. The space required for cable terminations may be around 2,000 m^2 in order to place a terminal tower (pylon) somewhat heavier in appearance than a normal suspension tower, a small building, and other transmission equipment.

AC underground lines by cable also have limitations on transmission distance due to the high capacitive reactance of cables. This problem is overcome by HVDC transmission. HVDC Light is a new underground power transmission and distribution system technology that is being developed for electricity distribution to remote areas. In the future, the powerline ampacity system methodology developed in this book will be extended to include underground cables and other current-carrying transmission and distribution line equipment.

---

* There were other papers on this subject at Cigre 1998.
** This discussion is based on a report on the World Wide Web by the National Grid Company (NGC), UK, 1998.

Dynamic Rating of Substation Equipment

The powerline rating system methodology may be easily adapted to determine the ampacity of other transmission and distribution line equipment. The calculation of equipment ampacity is much easier because, unlike a transmission line, it is situated at one location in a substation. For the calculation of equipment rating, a thermal model of the device is required along with a knowledge of the weather conditions at that location. Weather conditions are generally monitored in utility substations, enabling equipment ratings to be easily determined from this data. A simple equipment rating model (Douglass and Edris, 1996), (Soto et al., 1998) for implementation in the LINEAMPS program is given below:

$$I = I_r \cdot \frac{\left(\dfrac{T_{max} - T_a}{T_{max(r)} - T_{a(r)}} - \exp(-t/\tau) \right)}{1 - \exp(-t/\tau)}$$

I = Equipment ampacity at ambient temperature T_a
I_r = Nameplate rating of equipment
Tmax = Maximum equipment operating temperature
Tmax(r) = Rated maximum temperature of equipment specified for nameplate rating at rated ambient temperature Ta(r).
τ = Thermal time constant of substation equipment
t = time (for steady state rating t = ∞)
Ta = Ambient temperature (actual or forecast)
a = experimental constant

If device ratings limit line capacity it is relatively less expensive to replace a device with a higher rating device.

Line Maintenance

A line ampacity system having forecast capability is required for the planning of transmission line maintenance. When taking a line out of service to examine joints, inspect conductors, clean insulators, etc., it is necessary to ensure that adjacent transmission circuits will have sufficient line ampacity to safely carry the additional load of the line taken out of service.

Korea Electric Power Company (KEPCO) engineers used the LINEAMPS program (Choi et al., 1997) to safely re-conductor existing transmission circuits by replacing ACSR conductors with high-ampacity Zirconium aluminum alloy conductor and Invar core. Conductor replacement was carried out one circuit at a time on a double-circuit line by leaving the other circuit energized. For the planning of conductor replacement work they used LINEAMPS ratings to ensure that the energized circuit had sufficient capacity to carry the load of both circuits.

Summary, Future Plans and Conclusion

As transmission line ampacity is increased by dynamic line ratings, better techniques will be required to ensure that lines operate safely. Line maintenance using infrared imaging from a helicopter is becoming routine for many utilities (Burchette, Sam N., 1989). Similarly, devices for the robotic maintenance of conductors and line hardware are being developed so that long transmission circuits can be inspected rapidly without human intervention in areas of high electric and magnetic fields.

Deregulation, Independent System Operator and Power Pool Operations

The deregulation of the electric supply business is under active consideration worldwide (Cigré Panel Session, 1998), (Cigré Regional Meeting, 1996) and is already implemented in certain states (Barkovitch, B., Hawk, D.) of the U.S.A. and Canada (*Electricity Today*, 1998). Figure 11.1 is a simplified diagram of an unbundled system to illustrate the operation of a deregulated electric supply business. Independent generation companies operate power generation (G), distribution of electricity to consumers is done by independent distribution companies (D), and the transmission network is operated by a transmission company (T). An independent system operator (ISO) would ensure the reliability of the system as well as act as a clearinghouse for the purchase and sale of electricity through a competitive bidding process. In this scenario, a line ampacity system will provide assistance to the ISO in determining transmission line capacity and pricing for the transmission of power from (G) to (D).

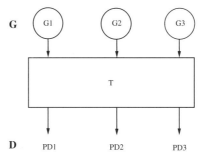

FIGURE 11.1 Operation of an Open Electricity Market.

According to (Cigré Panel Session, 1998), the concept of electricity deregulation is accepted almost universally, but the rules governing a competitive electric energy marketplace, ISO, and power pool operations are not yet well established. There is general consensus, however, that the rules should result in optimal system operations and encourage optimum system expansion plans. This is not an easy task and much remains to be done in this area. The application of the LINEAMPS program in the ideal operation of a power system as described in Chapter 8 is a contribution in this direction.

11.4 A PLAN TO DEVELOP LINEAMPS FOR AMERICA

Discussions with various electric power companies in the U.S. have revealed that many companies are still using a static line rating. The LINEAMPS program can be easily adapted for implementation in these utilities with minimum cost. Near real-time weather data is available from the National Weather Service (NWS). The NWS has an extensive network of weather stations connecting most regions of the U.S., and near real-time data is available on the Internet.* In addition, there are several private weather service companies (Figure 11.6) that deliver custom weather data. These data are adequate for LINEAMPS to determine powerline ampacity in all regions of the U.S. Similarly, Canadian and Mexican weather data is easily available from various agencies.

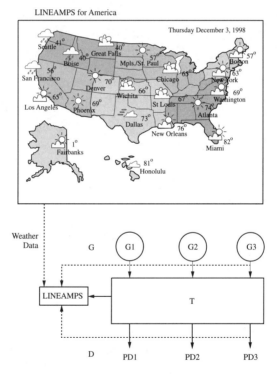

FIGURE 11.6 A proposal for LINEAMPS for North America (including USA, Canada and Mexico).

As deregulation of the electricity supply business is embraced by all states in the U.S., Canada, and even Mexico, it will enable the development of a more competitive business environment for the purchase and sale of electricity. In this scenario, determining transmission capacity will become more important. Advance knowledge of transmission capacity (hourly, daily, weekly, or longer) will enable

---

* US WEATHER SERVICE NOAA Internet Web Site: http://www.nws.noaa.gov/

Summary, Future Plans and Conclusion

more efficient electric energy futures trading by competitive bidding and spot pricing of electricity (Schweppe, 1987). All players in the electricity market can easily use the LINEAMPS program because it does not require any hardware connection with the transmission line or any other power system equipment.

A plan to develop LINEAMPS for America is shown in Figure 11.6. As shown in the figure, weather service data from different weather stations are input to LINEAMPS. The user of the program will simply select a transmission line of interest from the stored database of transmission lines to determine the ampacity of that line. In addition to forecast ratings, the program will also offer real-time dynamic and transient line ratings when real-time line current data is automatically entered to the program.

A real-time interface has already been developed by the author by establishing a DDE* connection with a Microsoft Excel spreadsheet. As seen in Figure 11.7, the program can be used by generation companies to determine the maximum generation capacity they can deliver through the transmission lines; by the transmission companies to determine transmission line capacity and determine transmission congestion and transmission pricing; and by distribution companies to determine the maximum generation capacity they can buy from low-cost generation supply sources. As of this writing, the author has contacted the South West Power Pool; OG&E, Oklahoma City; Houston Light & Power; Orange & Rockland Utilities, New York; PEPCO, Washington DC; PJM; and Virginia Power, Richmond, VA, for support in the implementation of this program. The implementation of this proposal will result in the development of a more efficient transmission network in America, enabling greater competition for the generation, transmission, and distribution of electricity with the objective of minimizing costs and lowering the price of electricity. In fact there is no reason why an electricity user should not be able to shop around in the electric energy market for the best rates, just like any other commodity.

11.5 CONCLUSION

In this chapter, the main contributions in the book are summarized and a plan is laid out for further research and development to maximize transmission capacity.

Electricity is the prime mover of modern society and its efficiency depends upon research and development of the electric power system. The electric power transmission network is one of the most important components of the electric power system that is responsible for the reliable operation of the interconnected system that comprises diverse generation sources, all connected together, to serve the load in the most economical manner.

As the demand for electricity grows, the need for higher transmission capacity will increase, requiring better planning, operation, and maintenance of lines. In the past, new transmission lines were constructed to meet the needs of the electricity supply industry. With public concern about environmental protection, the cost of land, and other economic and demographic factors, it is becoming difficult to build new lines, and new methods are required to maximize the utilization of existing lines. This is the object of the new Line Ampacity System developed in the book.

---

* LINEAMPS Dynamic Data Exchange (DDE) for Real Time Ampacity Calculations. Unpublished report. Anjan K. Deb 1998.

Bibliography

Ampère, André-Marie, 1827. Memoir on the Mathematical Theory of Electrodynamic Phenomena, Uniquely Deduced from Experience. Encyclopedia Britannica.

Alexandrov et al., 1998. Overhead line designing in view of environmental constraints. Compact overhead lines. Cigré, 1998, Paris.

Anand, V.P., Deb, Anjan K., and Mukherjee, P.K. 1984.Choice of conductors for 400kV transmission lines. Central Board of Irrigation and Power, New Delhi, India.

Barkovitch, B.D. Hawk. 1996. Starting a New Course in California. *IEEE Spectrum*, Vol. 33, July 1996, p26.

Bergen, Arthur R., 1986. *Power System Analysis,* Prentice Hall.

Black, W.Z. et al., 1987. Critical span analysis of overhead conductors. IEEE/PES Conference Paper Number 87 SM 560-6, San Francisco.

Black, W.Z. and W.R. Byrd, 1983. Real-time ampacity model for overhead lines. IEEE Transactions on Power Apparatus and Systems, Vol. PAS-102, No. 7.

Böhme et al., 1998. Overhead transmission lines: design aimed to reduce the permitting time. Cigre, 1998 paper Number 22/33/36-06.

Bonicel, J.P., Tatat, O., 1998. Aerial optical cables along electrical power lines. *REE*, No. 3, March 1998.

Booch, G., 1993. Object Oriented Design with Applications, The Benjamin/Cummings Publishing Co.

Borgard, L., 1999. Grid voltage support at your finger tips. Portable SMES units provide Wisconsin Public Service Corporation voltage and VAR support. *Transmission & Distribution*, Vol. 51, No. 9, October 1999.

Bousquet, Jean-Luc, Loreau, B., Bechet, D., and Delomel, J.C., 1997. Un nouveau conducteur compact homogène en alliage d'aluminium pour les lignes 400 kV. *REE* No.10, Novem- ber 1997.

Buckles, W., Hassenzahl, W.V., 2000. Superconducting Magnetic Energy Storage. IEEE Power Engineering Review, Volume 20, Number 5, p. 16–20, May 2000.

Burchette, Sam N., 1989. Helicopter maintenance on energized EHV transmission lines. *Transmission & Distribution*, Vol. 41, No. 11, November 1989.

CEA Report, 1980. Development for an accurate model of ACSR conductor for calculating sags at high temperatures-Part II: Sag-Tension Program-STESS.

Cibulka, L., Steeley, W.J., and Deb, A.K., 1992. Le système ATLAS de PG&E d'évaluation dynamique de la capacité thermique d'une ligne de transport. Conférence Internationale des Grands Réseaux Électriques, Paris.

Cigré, 1999. The use of weather predictions for transmission line thermal ratings. Electra No. 186, October 1999.

Cigré, 1999. The thermal behaviour of overhead conductors. Section 4. Mathematical model for evaluation of conductor temperature in the adiabatic state. Working Group 22.12.

_____1997. The thermal behaviour of overhead conductors. Section 3. Report prepared by Cigré Working Group 22.12.

_____1992. Section 1 and 2, Electra # 144.

_____ 1978. Permanent elongation of conductors predictor equations and evaluation methods. Cigré report # WG-05-22-78.

Cigré Panel Session, 1998. Impact of Deregulation on Operation and Planning of Large Power Systems.

Cigré Regional Meeting, 1996. Power Pool Arrangements and Economic Load Dispatch. New Delhi, India, 13–14 October 1996.

Cigré Symposium, 1991. Compacting Overhead Transmission Lines, Leningrad, USSR, 3–5 June 1991.

_____ 1985. High Currents in Power Systems under Normal, Emergency and Fault Conditions. Cigré Symposium, Brussels, Belgium, 3–5 June 1985.

Cox, Earl, 2000. What's the object? Bringing object oriented machine intelligence to web-hosted applications. PC AI, March/April 2000.

Davidson, Glen, 1969. Short-term thermal rating for bare overhead conductors. IEEE Vol. PAS-88, Number 3.

Davis, M.W. 1977. A new thermal rating approach: the real-time thermal rating system for strategic overhead conductor transmission lines. Part I: General description and justification of the real-time thermal rating system. IEEE Transaction PAS, Vol. 96.

Deb, Anjan Kumar, 1999. Object-oriented expert powerline ampacity system. US Patent Number 5,933,355 issued on August 3, 1999.

_____. 1998a. LINEAMPS: Line Ampacity System Comparison of Result with IEEE and Cigré. Contribution to Cigré 98, Paris, France.

_____. 1998b. A method for the calculation of transmission line conductor temperature and current from general heat equation. A Syntopicon Discussion on Descartes. Project IV, Columbia Pacific University, San Rafael, CA.

_____. 1998c. Application of neural network and fuzzy set theory. Response 6 CPU Course # EG 628: Ecology. January 23, 1998.

_____. 1995a. Object oriented expert system estimates power line ampacity. IEEE Computer Application in Power, Vol. 8, No. 3.

_____. 1995b. Object-oriented expert line ampacity system. Cigré Regional Meeting on Power Pool Arrangements and Economical Load Despatch. New Delhi, India.

_____. 1994. Thermal rating of overhead line conductor, SIEMENS, Electrizitwirtschaft, Germany

_____. 1993. Probabilistic design of transmission line sag and tension. International Conference on Electricity Distribution System, Birmingham, UK.

_____. 1982. Etude de la capacite thermique des conducteurs de lignes aeriennes. Memoire EDF/CNAM.

Deb, Anjan K., Singh, S.N., and Ghoshal,T.K. 1985. Higher service current in overhead lines. Cigré Brussels Symposium on High Currents in Electric Network under Normal and Emergency Conditions, Brussels.

Waterman, Donald A., 1986. A Guide to Expert Systems, Addison Wesley, Reading, MA.

Dorf, Richard C., 1993. *The Electrical Engineering Handbook*, CRC Press, Boca Raton.

Douglass, Dale A., 1986. Weather-dependent versus static thermal line ratings. IEEE Power Engineering Society, Transmission and Distribution Meeting, Anaheim, California, September 14-19, 1986, Paper 86 T&D 503-7. Discussion contribution of above paper by Anjan K. Deb.

Duffie, John A., Beckman, William A., 1980. *Solar Engineering of Thermal Processes*, John Wiley & Sons, New York, 1980.

Eberhart, R.C., Dobbins, R.W., 1990. *Neural Networks PC Tools – A Practical Guide*, Academic Press, New York.

Bibliography

Electricity Today, 1998. An Industrial Customer's Guide to an Open Electricity Market, Vol. 10, No. 8, September 1998.

Faraday, Michael, 1834. Experimental Researches in Electricity. Encyclopaedia Britannica.

Feak, S.D., 1997. Superconducting magnetic energy storage (SMES) utility application studies. IEEE PES SM 472-1 T-PWRS.

Foss, Stephen D., Maraio, Robert, 1989. Dynamic line rating in the operating environment. IEEE Transmission and Distribution Conference paper # 89 TD 431-8 PWRD.

Ghanoum, E., 1983. "Probabilistic Design of Transmission Lines", Part I, II. IEEE Transactions on Power Apparatus and Systems, Vol. PAS-102, No. 9, 1983.

Giacomo, D., Nicolini, G.P., and Paoli, P., 1979. Criteria for the statistical evaluation of the temperature in conductors, thermal rating problems. Report prepared by CIGRE working group 22. Cigré Sienna Colloquium. Italy.

Hall, J. F., Savoullis, J., and Deb, Anjan K., 1988. Wind tunnel studies of transmission line conductor temperature. IEEE Transactions on Power Delivery, Vol. 3, No. 4, pp. 801–812.

Hall, J.F. and Deb, Anjan K., 1988a. Economic evaluation of dynamic thermal rating by adaptive forecasting. IEEE Transactions on Power Delivery, Vol. 3, No. 4, pp. 2048–2055.

_____. 1988b. Prediction of overhead transmission line ampacity by stochastic and deterministic models. IEEE Transactions on Power Delivery, Vol. 3, No. 2.

Harvey, J. R., 1972. Effect of elevated temperature operation on the strength of aluminum conductors. IEEE Transactions, Power Engineering Society.

Haykin, S., 1999. *Neural Networks - A Comprehensive Foundation.* 2nd Edition, Prentice Hall, Englewood Cliffs, NJ.

Hingorani, N.G., Gyugyi, L., 2000. Understanding FACTS-Concepts and Technology of Flexible AC Transmission System. IEEE Press, 2000.

Hitachi, 1999. Wire catalog.

Howington, B.S. and Ramon, G.J., 1984. Dynamic thermal line rating, summary and status of the state-of-the-art technology. IEEE/PES 1984 Summer Meeting, Seattle.

House, P.D. and Tuttle, H., 1957. Current carrying capacity of ACSR. AIEE Transactions.

IEEE Power Engineering Review. 1998. Supergrids. A Win-Win Solution for Sustainable Development. Volume 18, Number 8, August 1998.

IEEE Std 738. 1993. IEEE Standard for calculating the current-temperature relationship of bare overhead conductors.

Jackson, R.L. and Price, C.F., 1985. Examination of continuous and thermal capacity of overhead lines. Cigré Brussels Symposium on High Currents in Electric Network under Normal and Emergency Conditions, Brussels.

Johannet, P., Dalle, B., 1979. Calcul des chutes de tension, des échauffements et des efforts électrodynamiques en cas de court-circuit. Lignes Aériennes. Techniques de l'ingénieur 12-1979.

Kappa-P.C., 1998. Object-Oriented Application Development Software, Version 2.4. Intellicorp, Mountain View California, 1998.

Kennedy, B. 2000. Artificial intelligence improves power quality. Electrical World Vol. 214, No. 3, May/June 2000.

Koch, B., 1999. Conductor developments help extend T&D systems. *Electrical World*, Vol. 213, No. 3, May/June 1999.

Koval, D.O., Billinton, R., 1970. Determination of transmission line ampacities by probability and numerical methods. IEEE Transactions on Power Apparatus and Systems, Vol. 89, No. 7.

Kronfeld, Kevin M., Tribble, Alan C., 1998. Expert System vs. Procedural Language Development. PC AI, July/August 1998.

Lankford, Craig B., 1989. ALCOA's Sag and Tension Program Enhanced for PC Use. Transmission and Distribution Journal, Vol 41, No 11, November 1989.

Stevens, Lawrence, 1993. *Artificial Intelligence. The Search for the Perfect Machine*. Hayden Book Company.

Lee, K.Y. et al., 1998. Adaptive Hopfield Neural Networks for Economical Load Dispatch. IEEE Transactions on Power Apparatus and Systems, Vol. 13, No. 2, May 1998.

Legrand, M.J.C., 1945. Les limites de puissancet des lignes électriques aériennes du fait de l'échauffement. Cigré.

Mauldin, T. Paul, Steeley, William J., Deb, A.K. 1988. Dynamic thermal rating of transmission lines independent of critical span analysis. International Conference on High Technology in the Power Industry, Phoenix. March 1–4.

Miner, Gayle F., 1996. *Lines and Electromagnetic Fields for Engineers*. Oxford University Press, New York.

Machowski, Jan, Bialek, Janusz W., Bumby, James R., 1998. *Power System Dynamics and Stability*, John Wiley & Sons.

Mizuno, Yukio, Nakamura, Hisahide, Adomah, Kwabena, Naito, Katsuhiko, 1998. Assessment of thermal deterioration of transmission line conductor by probabilistic method. *IEEE Transactions on Power Delivery*, Vol. 13, No. 1.

Mizuno, Yukio, Adomah, Kwabena, Naito, Katsuhiko, 2000. Probabilistic Assessment of the Reduction in Tensile Strength of an Overhead Line's Conductor with Reference to Climate Data. IEEE paper number PE015PRD (2-2000).

Morgan, Vincent T., 1991. Thermal behaviour of electrical conductors. Research Studies Press Ltd., Somerset, UK.

Morgan, V.T., 1978. The loss of tensile strength of hard-drawn conductors by annealing in service. *IEEE Transactions, Power Engineering Society*.

MPS Review Article. 1998a. New system improves IT functionality. *Modern Power System*, UK.

_____. 1998b. Generation without transformers: Introducing Powerformer. *Modern Power Systems*, UK.

Muhamed Aganagic, K.H. Abdul-Rahman, Waight, J.G., 1998. Spot pricing of capacities for generation and transmission of reserve in an extended Poolco model. *IEEE Transactions on Power Systems*, Vol. 13, No. 3, August 1998.

Nasar, S.A., Trutt, F.C., 1999. *Electric Power Systems*, CRC Press, Boca Raton.

National Electric Safety Code. 1997. C2-1997.

Negnevitsky, Michael, 1998. A knowledge-based tutoring system for teaching fault analysis. IEEE Transactions on Power Systems, Vol. 13, No. 1, February.

Noroozian, M., Ängquist, L., Ghandhari, M., Andersen, G., 1997. Use of UPFC for optimal power flow control. IEEE Transactions on Power Delivery, Vol. 12, No. 4.

Ozisik, M. Necati, 1985. Heat Transfer. McGraw Hill, New York.

Pfaffenberger, R.C., Patterson, J.H., *Statistical Methods for Business and Economics*, Richard D. Irwin, Inc.

PG&E Standard. 1978. Ampacity of Overhead Conductors.

PG&E Report. 1989. Risk Analysis of ATLAS at three PG&E locations. PG&E report 1989. Prepared by Anjan K. Deb with contributions from William J. Steeley (PG&E) and Benjamin L. Norris (PG&E).

Porcheron, Y., Hautefeuille, P., 1982. Lignes Aériennes. *Techniques de l'Ingénieur*, Paris.

Priestley, M.B., 1981. *Spectral Analysis and Time Series*, Volume 1, II. Academic Press, New York.

Rashkes, V.S., Lordan, R., 1998. Magnetic Field Reduction Methods: Efficiency and Costs. IEEE Transactions on Power Delivery, Vol. 13, No. 2.

Bibliography

Redding, J.L. 1993. A Method for Determining Probability Based Allowable Current Ratings For BPA's Transmission Lines. IEEE/PES 1993 Winter Meeting Conference Paper # 93WM 077-8PWRD, Columbus, Ohio, January 31–February 5, 1993.

Renchon, R., Daumerie, G., 1956. Image thermique de ligne aérienne. Cigré, (30 May–9 June).

Saadat, H., 1999. *Power System Analysis*, McGraw Hill, New York.

Schmidt, N.P., 1997. Comparison between IEEE and Cigré Ampacity Standards. IEEE PE-749-PWRD-0-06-1997. Discussion contribution by Anjan K. Deb.

Schweppe, Fred C. et al.,1988. Spot Pricing of Electricity. Kluwer Academic Publishers

Seppa, T.O., et al., 1998. Use of on-line tension monitoring for real-time thermal ratings, ice loads, and other environmental effects. Cigré 98 paper # 22–102.

Soto, F. et al., 1998. Increasing the capacity of overhead lines in the 400kV Spanish transmission network: real-time thermal ratings. Cigré '98 paper # 22-211.

Southwire, 1994. Overhead Conductor Manual. Southwire Company, Carrolton, GA.

Steeley, W.J., Norris, B.L., and Deb, A.K., 1991. Ambient temperature corrected dynamic transmission line ratings at two PG&E locations. IEEE Transactions on Power Delivery, Vol. 6, No. 3.

Subrahmanyam, V., 1996. *Power Electronics*, John Wiley & Sons, New York.

System for rating electric power transmission lines and equipment.1992. U.S. Patent 5,140,257.

Taylor, Odin, Smith, Peter, MacIntyre, John, Tait, John., 1998. Expert Systems for Power Industry. Handbook of Expert Systems. Editor: Jay Liebowitz. CRC Press, Boca Raton.

Thrash, R.F., 1999. Transmission conductors – A review of the design and selection criteria. Southwire Company.

Urbain, J.P., 1998. Intensité admissibles dans les conducteurs de lignes aériennes HTB lors des regimes temporaires de secours. EDF CNIR 98.

Waldorf, Stephen P., Engelhardt, John S., 1998. The first ten years of real-time ratings on underground transmission circuits, overhead lines, switchgear and power transformers. Electricity Today. Vol. 10, No. 2, February 1998.

Waterman, Donald A., 1987. *A Guide to Expert Systems*, Addison Wesley, Reading, MA.

Wong. K.P., Yuryevich., 1998. Evolutionary-Programming-Based Algorithm for Environmentally-Constrained Economic Dispatch. *IEEE Transactions on Power Delivery*, Vol. 13, No. 2, May 1998, p. 301–306.

Wood, Allen J., Wollenberg, Bruce F., 1996. *Power Generation Operation and Control,* John Wiley & Sons, New York.

Wook, M. Byong, Choi, Michael, Deb, Anjan K., 1997. Line rating system boosts economical energy transfer. IEEE Computer Application in Power, Vol. 8, No. 3.

Yalcinov, T., Short, M.J., 1998. Neural Networks for Solving Economic Dispatch Problem with Transmission Capacity Constraints. IEEE Transactions on Power Systems, Vol. 13, No. 2.

Appendix A

APPENDIX A.1
[1]US CONDUCTOR SIZES: ACSR/AS (ASTM B549-83)

Conductor	Construction				Area mm²		Diameter mm		Weight kg/km			RTS	Rdc 20°C
	Aluminun		Steel										
Code	No	Wire dia	No	Wire dia	Al	St	ACSR	St	ACSR	St	Al	kN	ohm/km
Kiwi	72	4.41	7	2.94	1097.71	47.46	44.07	8.82	3364	311	3053	215	0.0261
Bluebird	84	4.07	19	2.44	1091.75	88.87	44.76	12.20	3620	584	3036	264	0.026
Chukar	84	3.70	19	2.22	901.74	73.51	40.68	11.10	2991	483	2508	218	0.0314
Falcon	54	4.36	19	2.62	805.45	102.07	39.23	13.08	2910	671	2240	241	0.0347
Lapwing	45	4.78	7	3.18	805.43	55.67	38.20	9.55	2594	365	2229	186	0.0352
Bobolink	45	4.53	7	3.02	724.58	50.12	36.23	9.06	2334	329	2005	167	0.0391
Martin	54	4.02	19	2.41	684.36	86.63	36.16	12.05	2472	569	1903	204	0.0408
Pheasant	54	3.90	19	2.34	644.42	81.60	35.09	11.69	2328	536	1792	192	0.0433
Bittern	45	4.27	7	2.85	644.08	44.54	34.16	8.54	2074	292	1782	149	0.044
Skylark	36	4.78	1	4.78	644.35	17.90	33.42	4.78	1891	117	1775	115	0.0443
Grackle	54	3.77	19	2.27	603.76	76.58	33.97	11.33	2182	503	1679	181	0.0462
Bunting	45	4.14	7	2.76	603.99	41.74	33.08	8.27	1945	274	1672	139	0.0469
Finch	54	3.65	19	2.19	563.81	71.47	32.83	10.94	2037	469	1568	169	0.0495
Bluejay	45	3.99	7	2.66	563.79	39.00	31.96	7.99	1816	256	1560	130	0.0502
Curlew	54	3.51	7	3.51	523.14	67.81	31.62	10.54	1892	445	1448	161	0.0531
Ortolan	45	3.85	7	2.57	523.06	36.15	30.78	7.70	1685	237	1448	121	0.0541
Tanager	36	4.30	1	4.30	523.26	14.53	30.12	4.30	1536	95	1441	93	0.0546
Cardinal	54	3.38	7	3.38	483.13	62.63	30.38	10.13	1748	411	1337	149	0.0575
Rail	45	3.70	7	2.47	483.08	33.42	29.59	7.40	1556	219	1337	111	0.0586
Canary	54	3.28	7	3.28	455.77	59.08	29.51	9.84	1649	387	1261	140	0.0609
Mallard	30	4.14	19	2.48	402.66	91.88	28.95	12.41	1721	604	1117	172	0.067
Condor	54	3.08	7	3.08	402.39	52.16	27.73	9.24	1456	342	1114	124	0.069
Tern	45	3.38	7	2.25	402.61	27.82	27.01	6.75	1297	182	1114	93	0.0703
Coot	36	3.77	1	3.77	402.51	11.18	26.42	3.77	1182	73	1109	72	0.071
Drake	26	4.44	7	3.45	402.72	65.56	28.13	10.36	1544	430	1115	142	0.0682
Cuckoo	24	4.62	7	3.08	402.65	52.16	27.74	9.24	1456	342	1114	125	0.069
Redwing	30	3.92	19	2.35	362.25	82.51	27.45	11.76	1547	542	1005	154	0.0745
Starling	26	4.21	7	3.28	362.44	59.01	26.69	9.83	1390	387	1003	128	0.0758
Gannet	26	4.07	7	3.16	337.59	54.94	25.75	9.49	1294	360	934	119	0.0814
Flamingo	24	4.23	7	2.82	337.74	43.76	25.40	8.47	1222	287	935	105	0.0822
Swift	36	3.38	1	3.38	322.09	8.95	23.63	3.38	945	58	887	57	0.0887
Egret	30	3.70	19	2.22	322.05	73.51	25.89	11.10	1376	483	893	138	0.0838
Scoter	30	3.70	7	3.70	322.05	75.15	25.89	11.09	1386	493	893	143	0.0836
Grosbeak	26	3.97	7	3.09	322.17	52.43	25.16	9.27	1235	344	892	114	0.0853
Rook	24	4.14	7	2.76	322.13	41.74	24.81	8.27	1165	274	892	100	0.0862
Kingbird	16	4.78	1	4.78	286.38	17.90	24.81	8.27	910	117	793	65	0.0991
Teal	30	3.61	19	2.16	306.40	69.85	25.25	10.82	1309	459	850	131	0.0881
Wood Duck	30	3.61	7	3.61	306.40	71.49	25.25	10.82	1319	469	850	136	0.0879
Peacock	24	4.03	7	2.69	306.59	39.76	24.21	8.07	1109	261	848	95	0.0906
Eagle	30	3.46	7	3.46	281.77	65.75	24.21	10.38	1213	431	782	125	0.0956
Dove	26	3.72	7	2.89	281.83	45.93	23.54	8.67	1081	301	780	100	0.0975

[1] Conductor data is calculated by program based on individual wire properties given in Appendix B. For exact values refer to the standard.

APPENDIX A.2
²US CONDUCTOR SIZES: ACSR (ASTM B232-81)

Conductor Code	Construction Aluminun		Steel		Area mm²		Diameter mm		Weight kg/km			RTS	Rdc 20°C
	No	Wire dia	No	Wire dia	Al	St	ACSR	St	ACSR	St	Al	kN	ohm/km
Kiwi	72	4.41	7	2.94	1097.71	47.46	44.07	8.82	3423	371	3053	212	0.0263
Bluebird	84	4.07	19	2.44	1091.75	88.87	44.76	12.20	3732	696	3036	257	0.0263
Chukar	84	3.70	19	2.22	901.74	73.51	40.68	11.10	3083	575	2508	212	0.0319
Falcon	54	4.36	19	2.62	805.45	102.07	39.23	13.08	3039	799	2240	233	0.0355
Lapwing	45	4.78	7	3.18	805.43	55.67	38.20	9.55	2664	435	2229	181	0.0356
Bobolink	45	4.53	7	3.02	724.58	50.12	36.23	9.06	2397	391	2005	163	0.0396
Martin	54	4.02	19	2.41	684.36	86.63	36.16	12.05	2581	678	1903	198	0.0417
Pheasant	54	3.90	19	2.34	644.42	81.60	35.09	11.69	2431	639	1792	186	0.0443
Bittern	45	4.27	7	2.85	644.08	44.54	34.16	8.54	2130	348	1782	145	0.0445
Skylark	36	4.78	1	4.78	644.35	17.90	33.42	4.78	1914	139	1775	113	0.0446
Grackle	54	3.77	19	2.27	603.76	76.58	33.97	11.33	2278	599	1679	175	0.0473
Bunting	45	4.14	7	2.76	603.99	41.74	33.08	8.27	1998	326	1672	136	0.0475
Finch	54	3.65	19	2.19	563.81	71.47	32.83	10.94	2127	559	1568	163	0.0507
Bluejay	45	3.99	7	2.66	563.79	39.00	31.96	7.99	1865	305	1560	127	0.0509
Curlew	54	3.51	7	3.51	523.14	67.81	31.62	10.54	1978	530	1448	156	0.0543
Ortolan	45	3.85	7	2.57	523.06	36.15	30.78	7.70	1730	282	1448	118	0.0548
Tanager	36	4.30	1	4.30	523.26	14.53	30.12	4.30	1554	113	1441	92	0.0549
Cardinal	54	3.38	7	3.38	483.13	62.63	30.38	10.13	1826	489	1337	144	0.0588
Rail	45	3.70	7	2.47	483.08	33.42	29.59	7.40	1598	261	1337	109	0.0594
Catbird	36	4.14	1	4.14	483.20	13.42	28.95	4.14	1435	104	1331	85	0.0594
Canary	54	3.28	7	3.28	455.77	59.08	29.51	9.84	1723	461	1261	136	0.0623
Mallard	30	4.14	19	2.48	402.66	91.88	28.95	12.41	1836	719	1117	165	0.0697
Condor	54	3.08	7	3.08	402.39	52.16	27.73	9.24	1521	407	1114	120	0.0706
Tern	45	3.38	7	2.25	402.61	27.82	27.01	6.75	1332	217	1114	91	0.0712
Drake	26	4.44	7	3.45	402.72	65.56	28.13	10.36	1627	512	1115	137	0.0702
Cuckoo	24	4.62	7	3.08	402.65	52.16	27.74	9.24	1522	407	1114	121	0.0706
Redwing	30	3.92	19	2.35	362.25	82.51	27.45	11.76	1651	646	1005	148	0.0775
Starling	26	4.21	7	3.28	362.44	59.01	26.69	9.83	1464	461	1003	123	0.0780
Gannet	26	4.07	7	3.16	337.59	54.94	25.75	9.49	1363	429	934	115	0.0838
Flamingo	24	4.23	7	2.82	337.74	43.76	25.40	8.47	1277	342	935	102	0.0841
Swift	36	3.38	1	3.38	322.09	8.95	23.63	3.38	957	70	887	57	0.0891
Egret	30	3.70	19	2.22	322.05	73.51	25.89	11.10	1469	575	893	132	0.0872
Secret	30	3.70	7	3.70	322.05	75.15	25.89	11.09	1480	587	893	137	0.0871
Grosbeak	26	3.97	7	3.09	322.17	52.43	25.16	9.27	1301	410	892	110	0.0878
Kingbird	18	4.78	1	4.78	322.17	17.90	23.88	4.78	1027	139	887	68	0.0887
Teal	30	3.61	19	2.16	306.40	69.85	25.25	10.82	1397	547	850	125	0.0916
Wood Duck	30	3.61	7	3.61	306.40	71.49	25.25	10.82	1408	558	850	130	0.0916
Peacock	24	4.03	7	2.69	306.59	39.76	24.21	8.07	1159	311	848	92	0.0927
Eagle	30	3.46	7	3.46	281.77	65.75	24.21	10.38	1295	514	782	120	0.0996
Dove	26	3.72	7	2.89	281.83	45.93	23.54	8.67	1139	359	780	96	0.1003

² Conductor data is calculated by program based on individual wire properties given in Appendix B. For exact values refer to the standard.

APPENDIX A.3
³US CONDUCTOR SIZES: ACSR (ASTM B232-81)

Conductor Code	Construction Aluminun No	Wire dia	Steel No	Wire dia	Area mm² Al	St	Diameter mm ACSR	St	Weight kg/km ACSR	St	Al	RTS kN	Rdc 20°C ohm/km
Osprey	18	4.47	1	4.47	281.70	15.65	22.33	4.47	898	122	776	60	0.1015
Hen	30	3.20	7	3.20	241.60	56.37	22.42	9.61	1111	440	670	103	0.1161
Hawk	26	3.44	7	2.67	241.38	39.32	21.78	8.02	975	307	668	82	0.1171
Flicker	24	3.58	7	2.39	241.60	31.34	21.49	7.16	913	245	669	73	0.1176
Pelican	18	4.14	1	4.14	241.60	13.42	20.68	4.14	770	104	665	51	0.1183
Lark	30	2.92	7	2.92	201.35	46.98	20.47	8.77	926	367	559	85	0.1393
Ibis	26	3.14	7	2.44	201.11	32.74	19.88	7.32	812	256	557	68	0.1406
Brant	24	3.27	7	2.18	201.33	26.09	19.61	6.54	761	204	557	61	0.1411
Chickadee	18	3.77	1	3.77	201.25	11.18	18.87	3.77	641	87	554	43	0.1421
Oriole	30	2.69	7	2.69	170.41	39.76	18.83	8.07	783	311	473	72	0.1646
Linnet	26	2.89	7	2.24	170.23	27.69	18.29	6.73	687	216	471	58	0.1661
Merlin	18	3.47	1	3.47	170.33	9.46	17.36	3.47	543	74	469	36	0.1678
Ostrich	26	2.73	7	2.12	151.89	24.72	17.27	6.36	613	193	420	52	0.1862
Partridge	26	2.57	7	2.00	135.12	22.02	16.30	6.01	546	172	374	46	0.2093
Waxwing	18	3.09	1	3.09	135.00	7.50	15.45	3.09	430	58	372	29	0.2118
Penguin	6	4.77	1	4.77	107.17	17.86	14.31	4.77	433	139	294	37	0.2612
Pigeon	6	4.25	1	4.25	84.95	14.16	12.74	4.25	343	110	233	30	0.3295
Quail	6	3.78	1	3.78	67.37	11.23	11.35	3.78	272	87	185	24	0.4155
Raven	6	3.37	1	3.37	53.52	8.92	10.11	3.37	216	69	147	19	0.5229
Robin	6	3.00	1	3.00	42.39	7.07	9.00	3.00	171	55	116	15	0.6603
Sparate	7	2.47	1	3.30	33.63	8.54	8.25	3.30	159	66	92	15	0.8217
Sparrow	6	2.67	1	2.67	33.63	5.60	8.02	2.67	136	44	92	12	0.8323
Swanate	7	1.96	1	2.61	21.13	5.36	6.54	2.61	100	42	58	10	1.3079
Swan	6	2.12	1	2.12	21.13	3.52	6.35	2.12	85	27	58	7	1.3247
Turkey	6	1.68	1	1.68	13.28	2.21	5.04	1.68	54	17	36	5	2.1080
Cochin	12	3.37	7	3.37	107.05	62.44	16.85	10.11	784	488	296	91	0.2488
Brahma	16	2.86	19	2.48	102.95	91.88	18.14	12.41	1004	719	285	122	0.2481
Dorking	12	3.20	7	3.20	96.64	56.37	16.01	9.61	708	440	267	82	0.2756
Dotterel	12	3.08	7	3.08	89.59	52.26	15.42	9.25	656	408	248	76	0.2972
Guinea	12	2.92	7	2.92	80.54	46.98	14.62	8.77	590	367	223	68	0.3307
Leghorn	12	2.69	7	2.69	68.16	39.76	13.45	8.07	499	311	189	58	0.3907
Minorca	12	2.44	7	2.44	56.13	32.74	12.20	7.32	411	256	155	48	0.4745
Petrel	12	2.34	7	2.34	51.54	30.06	11.69	7.02	377	235	143	44	0.5167
Grouse	8	2.54	1	4.24	40.52	14.13	9.32	4.24	221	110	112	23	0.6761

³ Conductor data is calculated by program based on individual wire properties given in Appendix B. For exact values refer to the standard.

APPENDIX A.4
[4]BRITISH CONDUCTOR SIZES: ACSR (BS 215: Part 2: 1970)

Conductor Code	Construction Aluminun No	Construction Aluminun Wire dia	Construction Steel No	Construction Steel Wire dia	Area mm² Al	Area mm² St	Diameter mm ACSR	Diameter mm St	Weight kg/km ACSR	Weight kg/km St	Weight kg/km Al	RTS kN	Rdc 20°C ohm/km
Gopher	6	2.36	1	2.36	26.23	4.37	7.08	2.36	106	34	72	9.17	1.067
Weasel	6	2.59	1	2.59	31.60	5.27	7.77	2.59	128	41	87	11.05	0.8859
Ferret	6	3.00	1	3.00	42.39	7.07	9.00	3.00	171	55	116	14.82	0.6603
Rabbit	6	3.35	1	3.35	52.86	8.81	10.05	3.35	213	69	145	18.48	0.5295
Horse	12	2.79	7	2.79	73.33	42.77	13.95	8.37	537	334	203	62.11	0.3632
Dog	6	4.72	7	1.57	104.93	13.54	13.95	8.37	396	106	290	31.97	0.2708
Wolf	30	2.59	7	2.59	157.98	36.86	18.13	7.77	726	288	438	67.05	0.1776
Dingo	18	3.35	1	3.35	158.57	8.81	16.75	3.35	505	69	437	33.66	0.1803
Lynx	30	2.79	7	2.79	183.32	42.77	19.53	8.37	843	334	509	77.81	0.153
Caracal	18	3.61	1	3.61	184.14	10.23	18.05	3.61	587	80	507	39.08	0.1553
Panther	30	3.00	7	3.00	211.95	49.46	21.00	9.00	974	386	588	89.96	0.1324
Jaguar	18	3.86	1	3.86	210.53	11.70	19.30	3.86	671	91	580	44.68	0.1358
Zebra	54	3.18	7	3.18	428.66	55.57	28.62	9.54	1620	434	1186	127.54	0.0663
Fox	6	2.79	1	2.79	36.66	6.11	8.37	2.79	148	48	100	12.82	0.7634
Mink	6	3.66	1	3.66	63.09	10.52	10.98	3.66	255	82	173	22.06	0.4436
Skunk	12	2.59	7	2.59	63.19	36.86	12.95	7.77	463	288	175	53.52	0.4214
Beaver	6	3.99	1	3.99	74.98	12.50	11.97	3.99	303	97	205	26.21	0.3733
Raccoon	6	4.09	1	4.09	78.79	13.13	12.27	4.09	318	102	216	27.54	0.3552
Otter	6	4.22	1	4.22	83.88	13.98	12.66	4.22	339	109	230	29.32	0.3337
Cat	6	4.50	1	4.50	95.38	15.90	13.50	4.50	385	124	261	33.34	0.2935
Hare	6	4.72	1	4.72	104.93	17.49	14.16	4.72	424	136	288	36.68	0.2667
Hyena	7	4.39	7	1.93	105.90	20.47	14.16	4.72	449	159	290	40.39	0.2633
Leopard	6	5.28	7	1.75	131.31	16.83	14.16	4.72	491	131	360	39.87	0.2144
Tiger	30	2.36	7	2.36	131.16	30.60	16.52	7.08	603	239	364	55.67	0.2139
Coyote	26	2.54	7	1.91	131.68	20.05	15.89	5.73	521	157	364	43.14	0.2151
Lion	30	3.18	7	3.18	238.15	55.57	22.26	9.54	1095	434	661	101.08	0.1178
Bear	30	3.30	5	7.00	256.46	192.33	22.26	9.54	2214	1502	711	267.01	0.1017
Batang	18	4.78	7	1.68	322.85	15.51	22.26	9.54	1017	121	896	65.62	0.0893
Goat	30	3.71	7	3.71	324.14	75.63	25.97	11.13	1490	591	899	137.59	0.0865
Antelope	54	2.97	7	2.97	373.92	48.47	26.73	8.91	1413	379	1035	111.25	0.076
Sheep	30	3.99	7	3.99	374.92	87.48	27.93	11.97	1723	683	1040	159.14	0.0748
Bison	54	3.00	7	3.00	381.51	49.46	27.00	9.00	1442	386	1056	113.51	0.0745
Deer	30	4.27	7	4.27	429.38	100.19	29.89	12.81	1974	783	1191	182.26	0.0653
Camel	54	3.30	5	7.00	461.63	192.33	29.89	12.81	2783	1502	1281	296.95	0.0592
Elk	30	4.50	7	4.50	476.89	111.27	31.50	13.50	2192	869	1323	202.42	0.0588
Moose	54	3.53	7	3.50	528.22	67.31	31.68	10.50	1988	526	1462	155.78	0.0538

[4] Conductor data is calculated by program based on individual wire properties given in Appendix B. For exact values refer to the standard.

Appendix A

APPENDIX A.5
[5]GERMAN CONDUCTOR SIZES: ACSR (DIN 48204-1974)

Conductor Size	Construction Aluminun No	Construction Aluminun Wire dia	Construction Steel No	Construction Steel Wire dia	Area mm² Al	Area mm² St	Diameter mm ACSR	Diameter mm St	Weight kg/km ACSR	Weight kg/km St	Weight kg/km Al	RTS kN	Rdc 20°C ohm/km
16/2.5	6	1.80	1	1.80	15.26	2.54	5.40	1.80	62	20	42	5.3	1.8341
25/4	6	2.25	1	2.25	23.84	3.97	6.75	2.25	96	31	65	8.3	1.1738
35/6	6	2.70	1	2.70	34.34	5.72	8.10	2.70	139	45	94	12.0	0.8152
44/32	14	2.00	7	2.40	43.96	31.65	8.10	2.70	367	246	120	44.4	0.5891
50/8	6	3.20	1	3.20	48.23	8.04	9.60	3.20	195	63	132	16.9	0.5803
50/30	12	2.33	7	2.33	51.14	29.83	11.65	6.99	375	233	142	43.3	0.5207
70/12	26	1.85	7	1.44	69.85	11.39	11.72	4.32	282	89	193	23.8	0.4048
95/15	26	2.15	7	1.67	94.35	15.33	13.61	5.01	381	120	261	32.1	0.2997
95/55	12	3.20	7	3.20	96.46	56.27	16.00	9.60	706	440	267	81.7	0.2761
105/75	14	3.10	19	2.25	105.61	75.51	16.00	9.60	882	590	292	106.0	0.2476
120/20	26	2.44	7	1.90	121.51	19.84	15.46	5.70	491	155	336	41.4	0.2327
120/70	12	3.60	7	3.60	122.08	71.22	18.00	10.80	894	556	338	103.4	0.2181
125/30	30	2.33	7	2.33	127.85	29.83	16.31	6.99	588	233	355	54.3	0.2194
150/25	26	2.70	7	2.10	148.79	24.23	17.10	6.30	601	189	412	50.6	0.1900
170/40	30	2.70	7	2.70	171.68	40.06	18.90	8.10	789	313	476	72.9	0.1634
185/30	26	3.00	7	2.33	183.69	29.83	18.99	6.99	741	233	508	62.4	0.1539
210/35	26	3.20	7	2.49	209.00	34.07	20.27	7.47	845	266	578	71.2	0.1353
210/50	30	3.00	7	3.00	211.95	49.46	21.00	9.00	974	386	588	90.0	0.1324
230/30	24	3.50	7	2.33	230.79	29.83	20.99	6.99	872	233	639	69.3	0.1231
240/40	26	3.45	7	2.68	242.93	39.47	21.84	8.04	981	308	672	82.6	0.1164
265/35	24	3.74	7	2.49	263.53	34.07	22.43	7.47	995	266	729	79.1	0.1078
300/50	26	3.86	7	3.00	304.10	49.46	24.44	9.00	1228	386	842	103.4	0.0930
305/40	54	2.68	7	2.68	304.46	39.47	24.12	8.04	1151	308	843	90.6	0.0933
340/30	48	3.00	7	2.33	339.12	29.83	24.99	6.99	1172	233	939	84.0	0.0843
380/50	54	3.00	7	3.20	381.51	56.27	27.60	9.60	1495	440	1056	121.6	0.0743
385/35	48	3.20	7	2.49	385.84	34.07	26.67	7.47	1334	266	1068	95.8	0.0741
435/55	54	3.20	7	3.20	434.07	56.27	28.80	9.60	1641	440	1201	129.2	0.0655
450/40	48	3.45	7	2.68	448.49	39.47	28.74	8.04	1549	308	1241	111.2	0.0638
490/65	54	3.40	7	3.40	490.03	63.52	30.60	10.20	1852	496	1356	145.8	0.0580
495/35	45	3.74	7	2.49	494.11	34.07	29.91	7.47	1634	266	1367	111.2	0.0580
510/45	48	3.68	7	2.87	510.28	45.26	30.69	8.61	1766	354	1412	126.9	0.0560
550/70	54	3.60	7	3.60	549.37	71.22	32.40	10.80	2077	556	1520	163.5	0.0517
560/50	48	3.86	7	3.00	561.42	49.46	32.16	9.00	1940	386	1554	139.2	0.0509
570/40	45	4.02	7	2.68	570.87	39.47	32.16	8.04	1888	308	1580	128.6	0.0502
650/45	45	4.30	7	2.87	653.16	45.26	34.41	8.61	2161	354	1808	147.3	0.0439
680/85	54	4.00	19	2.40	678.24	85.91	36.00	12.00	2559	672	1886	196.2	0.0421
1,045/45	72	4.30	7	2.87	1045.05	45.26	43.01	8.61	3260	354	2906	201.6	0.0277

[5] Conductor data is calculated by program based on individual wire properties given in Appendix B. For exact values refer to the standard.

APPENDIX A.6
⁶FRENCH STANDARD SIZES: AAAC (NF C 34-120 1976)

Conductor Size, mm²	Construction AAAC		Area mm²	Diameter mm	Weight kg.km	RTS kN	Rdc 20°C ohm/km
	Wires	Wire Dia, mm					
22	7	2.00	21.99	6	59.4	7	1.4793
34	7	2.50	34.36	8	92.9	11	0.9469
55	7	3.15	54.55	9	147.5	17	0.5966
76	19	2.25	75.55	11	204.4	23	0.4311
117	19	2.80	116.99	14	316.7	36	0.2785
148	19	3.15	148.07	16	401.0	46	0.2201
182	37	2.50	181.62	18	492.2	56	0.1796
228	37	2.80	227.83	20	617.7	70	0.1432
288	37	3.15	288.35	22	782.2	89	0.1132
366	37	3.55	366.23	25	994.1	113	0.0892
570	61	3.45	570.24	31	1550.4	176	0.0574
851	91	3.45	850.69	38	2317.8	262	0.0386
1144	91	4.00	1143.54	44	3120.1	347	0.0287
1596	127	4.00	1595.93	52	4363.9	484	0.0206

⁶ Conductor data is calculated by program based on individual wire properties given in Appendix B. For exact values refer to the standard.

APPENDIX A.7
⁷FRENCH STANDARD SIZES: ACSR (NF C 34-120 1976)

Conductor Size	Construction				Area mm²		Diameter mm		Weight kg/km			RTS kN	Rdc 20°C ohm/km
	Aluminun		Steel										
	No	Wire dia	No	Wire dia	Al	St	ACSR	St	ACSR	St	Al		
38	9	2.00	3	2.00	28.26	9.42	6.30	2.30	151	73	77	15.5	0.9669
60	12	2.00	7	2.00	37.68	21.98	10.00	6.00	276	172	104	31.91	0.7068
76	12	2.25	7	2.25	47.69	27.82	11.25	6.75	349	217	132	40.39	0.5584
116	30	2.00	7	2.00	94.20	21.98	14.00	6.00	433	172	261	39.98	0.2978
116	30	2.00	7	2.00	94.20	21.98	14.00	6.00	433	172	261	39.98	0.2978
147	30	2.25	7	2.25	119.22	27.82	15.75	6.75	548	217	331	50.61	0.2353
147	30	2.25	7	2.25	119.22	27.82	15.75	6.75	548	217	331	50.61	0.2353
182	30	2.50	7	2.50	147.19	34.34	17.50	7.50	677	268	408	62.48	0.1906
182	30	2.50	7	2.50	147.19	34.34	17.50	7.50	677	268	408	62.48	0.1906
228	30	2.80	7	2.80	184.63	43.08	19.60	8.40	849	337	512	78.37	0.1519
228	30	2.80	7	2.80	184.63	43.08	19.60	8.40	849	337	512	78.37	0.1519
288	30	3.15	7	3.15	233.67	54.52	22.05	9.45	1074	426	648	99.19	0.1201
288	30	3.15	7	3.15	233.67	54.52	22.05	9.45	1074	426	648	99.19	0.1201
297	36	2.80	19	2.25	221.56	75.51	22.45	11.25	1206	591	615	119.65	0.1247
412	32	3.60	19	2.40	325.56	85.91	26.40	12.00	1576	672	903	146.86	0.0858
612	42	2.61	19	2.65	224.59	104.74	23.69	13.25	1443	820	623	153.9	0.1208
865	66	3.72	19	3.15	716.97	147.99	30.63	15.75	3147	1158	1989	275.77	0.0393
1185	66	3.47	37	2.80	623.84	227.71	33.48	19.60	3513	1782	1731	354.37	0.0441

⁷ Conductor data is calculated by program based on individual wire properties given in Appendix B. For exact values refer to the standard.

Appendix A

HIGH AMPACITY CONDUCTOR COMPARED TO ACSR (A1.8)

Conductor	ACSR				High Ampacity Conductor			
	240mm²	330mm²	410mm²	480mm²	240mm²	330mm²	410mm²	480mm²
Construction	30/3.2	26/4.0	26/4.5	45/3.7	30/3.2	26/4.0	26/4.5	45/3.7
Al. alloy Invar steel	7/3.2	7/3.1	7/3.5	7/2.47	7/3.2	7/3.1	7/3.5	7/2.47
Rated tensile strength, kgf	10,210	10,930	13,890	11,800	9,170	10,000	12,720	10,500
Weight, kg/km	1100	1320	1,673	1,699	1,122	1,330	1,687	1,611
dc resistance, ohm/km	0.1200	0.0888	0.0702	0.06994	0.1220	0.0904	0.0714	0.0607
Modulus of elasticity, kg/mm²	9,081	8,346	8,368	7,253	8,231[1] 16,500[2]	7,720[1] 16,500[2]	7,730[1] 16,500[2]	6,960[1] 16,500[2]
expansion coefficient, 10-7/°C	17.97	18.97	18.95	20.84	15.3[1] 3.6[2]	17.0[1] 3.6[2]	17.0[1] 3.6[2]	19.8[1] 3.6[2]
Ampacity, A	595	720	850	917	1,173	1,425	1,675	1,810

Courtesy KEPCO
[1]Normal temperature,
[2]High temperature

Appendix B

WIRE PROPERTIES (B.1)

Wire Type	Copper	HC Al	HD Al EC	AA 6201	AA Almelec	AA Zr	AS 20	AS 40	HDG Steel
Conductivity %	100	62	61	52.5	51	60	20	40	9
Tensile Strength kg/mm^2	200	16	16	33	31.9	16	135	70	126.8
Max Temp Steady °C	80	80	80	80	80	200	90	90	300
Max Temp Dynamic °C	100	100	100	100	100	230	120	120	300
Max Temp Transient °C	180	180	180	180	180	300	200	200	300
Coeff. of linear expn. /°C × 10^{-3}	16.9	23	23	23	23	23	12.6	15.5	11
Density, kg/m^3	8890	2700	2700	2700	2700	2700	6530	4800	7780
Modulus of elasticity, kg/mm^2	11900	11900	6300	6700	6700	6700	15800	11100	20000
Specific heat, J/(kg K)	383	897	897	909	909	909	518	630	481
Temp. coeff of specific heat /°C	0.000335	0.00038	0.00038	0.00045	0.00045	0.00045	0.00014		0.0001
Temp. coeff of Rdc /°C	0.00393	0.004	0.004	0.0036	0.0036	0.0036	0.0036	0.0036	0.00327

Conductivity is based on % IACS (International Annealed Copper Standard)
Cu (100% IACS) = 0.017241 ohm.mm^2/m at 20°C
HC AL = High conductivity aluminum
HD AL = Hard drawn High aluminum (EC grade)
AA 6201 = Aluminum alloy (Al, Mg, Si) 6201 ASTM Standard
AA Almelec = Aluminum alloy (Al, Mg, Si) French Standard
AA Zr = Aluminum Zirconium alloy (Thermo resistant wire)
AS 20 = Aluminum clad steel wire (20% conductivity)
AS 40 = Aluminum clad steel wire (40% conductivity)
HDG Steel = Galvanized steel

Index

A

AAC (All Aluminum Conductors), 5
AACSR (Aluminum Alloy Conductor Steel Reinforced), 5
AC resistance of ACSR conductor, 30, 32, 51–59
ACAR (Aluminum Conductor Alloy Reinforced), 7
ACSR (Aluminum Conductor Steel Reinforced), 6, 30, 32, 51–59
ACSR conductor ampacity calculation, 30
Active shielding of transmission lines, 103
All Aluminum Alloy conductors, 5–8
Aluminum conductors, 5–8
Ambient temperature in ampacity calculations, 16
AmbientGen, 116, 139, 150, 215
Ampacity
    calculations, 15–16
    defining problem, 16–17
    effect of elevated temperature, 73
    substation equipment, 224
    transmission line, 219
Anode, 164
Applications, 183
    economic operation, 183
        electricity generation cost, 186–190
        formulation of optimization problem, 184–186
        long distance transmission, 198–201
        protection, 201–204
        stability of generators, 190
            dynamic stability, 191–193
            transient stability, 193–195
        transmission system planning, 195–198
ARMAV (Auto Regressive Moving Average) process, 125
Artificial neural network model, 127–131, 216
Average conductor temperature, 29

B

Back propagation algorithm, 130
Box-Jenkins forecasting model, 215
Bus voltage limits, 185

C

Capacity limits, 185
Cartograph object in computer model, 152–153
Cathode, 164
Cigré
    LINEAMPS comparison with standard, 66–71
    long distance electricity transportation, 221–222
    standard, 15
Circular flux AC resistance of ACSR conductor, 53–54
Coefficient of heat transfer, 37–39, 42
Coefficient of linear expansion, 79
Communication by fiber optics, 8, 23, 222
Complex voltage series, 174
Computer modeling, 143
    expert system design, 154
        expert system rules, 157–158
        goal-oriented programming, 155
    LINEAMPS expert system, 143
    object model, 144
        cartograph object, 152–153
        conductor object, 150–152
        LINEAMPS object model, 144
        transmission line object, 145–148
        weather station object, 148–150
    program description, 159
        LINEAMPS control panel, 159–161
        LINEAMPS windows, 159
        modeling transmission line and environment, 159
Computer programs
    LINEAMPS (see LINEAMPS program)
    required for prediction of sag and tension, 77–78
Conductivity, thermal, of transmission line, 28
Conductor configurations, 100–101
Conductor elongation. *see* Elongation of conductors
Conductor heat balance equation, 19
Conductor object in computer model, 150–152
Conductor sag. *see* Sag
Conductor temperature
    differential equation, 19–21, 25, 29
    frequency distribution, 87, 213
    measurement, 71
    probability distribution, 62, 73–75

245

use in ampacity calculations, 16
Conductor tension. *see* Tension
Conductor thermal modeling
   differential equation of conductor temperature, 19–21, 25, 29
   dynamic ampacity, 36–44
   general heat equation, 28–29
   radial conductor temperature, 47–49, 61, 212
   steady-state ampacity, 29–36
   transient ampacity, 44–47
   wind tunnel experiments, 61, 212
Conductor thermal ratings comparisons, 67–71
Conductors, 5–8, 11–12
Control center, 9
Control panel in LINEAMPS, 159–161
Cooling of transmission line, 62, 69–71
Cost savings, 186–190
Cost-benefit analysis, dynamic line rating, 12
Costs of transmission lines, 195–198
Creep, 81–82, 88
Current
   calculation of, 104, 205
   line current modeling, 215–217
Custom power, 178–179

D

DC convertors, 167
DC resistance of ACSR conductor, 32
Deregulation of electric supply business, 225
Design of expert system, 154–158
Deterministic model, time series, 24
Differential equation of conductor temperature, 19–21, 25, 29
Differential relays, 201
Diode, 164
Distributed temperature sensor system, 23–25
Distribution substation, 8–9, 223
Dynamic ampacity, 20–21, 36–44, 71
Dynamic conductor temperature, 36–44
Dynamic line rating systems
   electricity generation costs, 188
   operational cost savings, 12
   short-term rating, 212
Dynamic line ratings, 18, 19–21
Dynamic stability of generators, 191–193
Dynamic state operating conditions, 27

E

Earth fault relays, 201
Economic operation, 183
   electricity generation cost, 186–190
   formulation of optimization problem, 184–186
EHV (extra high voltage) network applications, 181
EHV transmission lines, 183, 198, 221–222
Electric field of transmission line, 93, 108–113, 219
Electric power system overview, 3–9
Electricity generation costs, 186
   cost savings by LINEAMPS, 188–190
   using LINEAMPS rating, 188
   using static line rating system, 186–188
Elevated temperature effects, 73
   change of state equation, 78–80
   elongation of conductor, permanent, 81–83
   loss of strength, 83–84
   transmission line sag and tension, 74–79
Elongation of conductors
   caused by elevated temperature, 11, 77, 79–80, 218
   geometric settlement, 81
   metallurgical creep, 81, 88
   predicted by probability modeling, 23
   recursive estimation of, 82
EMF mitigation measures, 102–103, 223
Energy control center, 9
Energy sources, renewable, 221–222
Environmental impact, 11, 220, 223
Equations
   allocation of load demand, 184
   ARMAV process in real-time forecasting, **125**
   back propagation algorithm, 130–131
   capital cost of line, 196–197
   change of state equation, 63–65, 78–79, **213**
   circular flux, 53–54
   conductor heat balance equation, 19
   conductor sag, 22, 63–66, 78–79
   conductor temperature differential equation, 20–21, 25, 29
   conductor tension, 63–66, 213, 219
   current through conductor, 94
   dynamic conductor temperature equation, **37**
   electric field, 109–110
   equipment rating model, 224
   forced convection cooling in conductor, **62**
   Fourier series weather model, 116–117
   fuzzy set theory for weather model, **132**
   general heat equation, 28
   geometric settlement, 81
   inelastic elongation, 82
   Kalman filter algorithm, 126–127
   Kohonen's learning algorithm, 131
   line ampacity problem definition, 16–17
   load current on thyristor, 165–166
   long-distance transmission, 198–199
   longitudinal flux, 51–53

Index

loss of tensile strength, 83–84
magnetic field inside conductor, 96–97
magnetic field of three-phase powerline, 98
magnetic field outside conductor, 95–96
Maxwell's, 94, 223
metallurgical creep, 81–82
optimum size of conductor, 197
reactive power, 170–171
real-time dynamic ampacity, 20–21
solar radiation, 137–139
specific transmission cost, 196
spectral analysis weather model, 117
stochastic model, 24
temperature of conductor, 19–20
time series stochastic and deterministic model, 75
transient conductor temperature, 44–45
transmission line, 205–207
voltage drop per meter, 54–55
Equipment ampacity, 224
Examples
   ACSR conductor
      dynamic ampacity, 43–44
      resistance, 55–59
      steady-state ampacity, 32–35
      temperature, 35–36, 41–43, 45–46
   electric field, 110–113
   electricity generation costs, 186
   fuzzy set calculation of ampacity, 135–137
   generator stability, 191, 192–193, 194
   long-distance transmission, 199–201
   magnetic field, 98–100
   magnetic field with shielding, 103–108
   New Zealand weather station object, 149–150
   object creation in LINEAMPS, 145–148
   rules used by LINEAMPS, 157–158
   series compensation, 174–178
Experimental verification of ampacity, 61
   conductor temperature measurement, 71
   IEEE and Cigré compared to LINEAMPS, 66–71
   outdoor test span experiment, 63–66
   wind tunnel experiments, 61–63
Expert line rating system, 25–26
Expert system design, 153–158
Expert systems, 9, 218

F

FACTS
   applications list, 180
   future R&D, 181, 220
   increasing transmission capacity (see Increasing transmission capacity, FACTS)
   manufacturers, 180
   recent developments, 12
   semiconductor valve assembly, 167–168
Fault discrimination, 201–202
Faults, importance of clearing fast, 47
Fiberoptic cable, 8, 23, 222
Fiberoptic conductor temperature sensors, 23
Firing angle of thyristor, 170
Flexible AC Transmission System. *see* FACTS
Forecasting models, 216
Fossil fuels, 221, 222
Fourier series weather model, 116–123, 216
Fuzzy set weather model, 132–137

G

Gate, 164
GBTR (giant bipolar transistor), 167
General heat equation, 28–29
Generator capacity limits, 185
Generator rotor oscillations, 191–192, 193
Generator, Powerformer, 12
Geometric settlement, 81
Geometries of transmission lines, 100–101
Global warming, 221
Goal-oriented programming, 155
GTO (gate turn-off) thyristor, 166

H

Heat effects on conductors, 11
Heat exchange during adiabatic condition, 45
Heat transfer coefficient, 37–39, 42
Helicopter, 224
High-voltage substation, 8–9
HV network applications, 180
HVDC (high-voltage DC transmission), 168–169, 223
HVDC LIGHT, 181, 223
Hydroelectric reserves, 221

I

Ice loads, 22
IEEE standard, 15, 66–71
IGBT (insulated gate bipolar transistor), 166–167
Impedance relays, 201
Increasing transmission capacity, 163
   FACTS, 168
      custom power, 178–179
      future R&D, 180–181
      HVDC, 168–181
      list of applications, 180
      manufacturers, 180

series compensation, 173–178
SMES, 179–180
STATCOM, 171–173
static VAR compensator, 169–171
UPFC, 178
New power semiconductor devices, 163
FACTS semiconductor valve, 167–168
GBTR, 167
GTO, 166
IGBT, 167
MCT, 167
MOSFET, 166
Thyristor, 16, 164–166
Independent system operators, 225
Infrared imaging, 225
Inheritance of objects, 144, 153
Integrated line ampacity system, 25–26
Invar alloy reinforced steel wires, 8, 224

K

Kalman filter algorithm, 126–127, 216
Kappa-PC, 144, 155
Kohonen's learning algorithm, 131

L

Learning in neural network, 129–130
Line ampacity problem definition, 16–17
Line maintenance, 224–225
Line rating methods, 15
 distributed temperature sensor system, 23–25
 dynamic line ratings, 17–18, 20–21
 expert line rating system, 25
 historical background, 15–16
 integrated line ampacity system, 25–26
 line ampacity problem, 16–17
 object-oriented modeling, 25
 online temperature monitoring system, 19–21
 online tension monitoring system, 21–22
 rating systems (see Rating transmission lines)
 sag monitoring system, 22–23
 static line ratings, 17
 weather-dependent systems, 18–19
Line security analysis, 74
LINEAMPS program
 computer modeling (see Computer modeling)
 conductor temperature
  dynamic state, 41
  steady-state, 32
  transient state, 47
 definition, 143–144
 evaluated by power companies, 211
 forecasting powerline ampacity, 18–19, 124

generating weather data, 24
IEEE and Cigré comparison, 66–71
outdoor test span comparison, 63–65
overview, 25
plan to develop in America, 226–227
sag calculation, 63–66
summary, 209–211
temperature measurement of transmission line, 71
tension calculation, 63–66
weather modeling, 116
wind tunnel data comparison, 62
LINEAMPS ratings, 188–190
Long distance transmission, 198–201
Longitudinal flux, AC resistance of ACSR conductor, 51–53
Loss of strength
 calculations for AAC Bluebell, 88
 caused by elevated temperature, 11, 75, 77, 79, 219
 determined experimentally, 65
 percentile method to calculate, 83
 predicted by probability modeling, 23
 recursive estimation of, 83–84

M

Magnetic field impact on environment, 11, 219, 223
Magnetic field of transmission line, 93
 conductor magnetic field, 94–95
  field inside conductor, 96–97
  field outside conductor, 95–96, 97
 different geometry, 100–102
 EMF mitigation, 102–108
 three-phase powerline magnetic field, 98–100
Manufacturers of FACTS devices, 180
Mathcad Solver, 58, 111
MCT (MOS-controlled thyristor), 167
Metallurgical creep, 81–82, 88
Meteorological conditions. *see* Weather conditions
Methods in computer program, 144
MOSFET, 166

N

National Weather Service
 data supplied, 116
 forecasts used in LINEAMPS program, 24–25, 148, 149–150, 226
Neural network weather model, 127–131, 216

Index

O

Object-oriented modeling, 25–26, 217
Objects in computer program, 144
Offline line ratings, 18, 22
Online line ratings, 18
Online temperature monitoring system, 19–21
Online tension monitoring system, 21–22
Operating condition of transmission line, 27
OPGW (Optical Ground Wire) conductor, 8, 23
Oscillations of generator rotor, 191–192, 193
Outdoor test span, 63–66
Overcurrent relays, 201
Overhead conductors, thermal ratings, 67–71
Overhead transmission lines, 4–8, 74
Overload protection, 201–204

P

Passive shielding of transmission lines, 102
Passive shunt capacitors and reactors, 169
Power balance equation, 184
Power companies using LINEAMPS, 211
Power Donut temperature sensors, 19
Power pool operations, 225
Powerformer generator, new technology, 12, 221
Powerline communication system, 8, 23
Pretensioning, 81
Probabilistic design of overhead lines, 74
Probabilistic rating methods, 23–25
Probability distribution of conductor temperature, 62, 73–75
Probability modeling of conductor temperature, 23
Program description of LINEAMPS, 159–161
Protection, 201–204

R

Radial conductor temperature, 47–49, 61, 212
Rating transmission lines
    dynamic, 12, 18, 188–189, 212
    LINEAMPS, 188–190
    static, 17, 186–188
    transient, 212
Real-time weather forecasting, 124–127, 213
Recursive estimation, 25, 216
Renewable energy sources, 221–222
Robotic maintenance, 225
Rotor angle, 191
Rotor oscillations, 191–192, 193
Rules used in LINEAMPS, 157–158

S

Sag
    calculation comparisons, 89–92
    calculation of, 23, 63–66, 78–79, 219
    caused by elevated temperature, 74–78, 83, 219
    effect on electric and magnetic fields, 93, 98, 108
    monitoring of, 18, 22
    problem, 75
Security of transmission line, 74
Semiconductors, 163–168
Series compensation, 173–178
Shielding of transmission lines, 102–103
Shunt and series capacitors, 221
SMES (Superconducting Magnetic Energy Storage), 179–180, 221
Solar energy development, 221
Solar radiation, 16, 137–139
SolarGen, 139, 149, 215
SSAC (Steel Supported Aluminum Conductor), 8
SSSC (Static Synchronous Series Compensator), 174
Stability of generators, 190
    dynamic stability, 191–193
    transient stability, 193–195
STATCOM (Static Synchronous Compensator), 171–173
Static line ratings, 17, 186–188
Steady-state ampacity, 29–36, 66–71
Steady-state line rating, 212
Steady-state operating conditions, 27
Steady-state stability of generators, 190–191
Stochastic model, time series, 23–25
Strength, loss of. *see* Loss of strength
Stress/strain relationship, 80
Substation, electric, 8–9, 224
Supervised learning in neural network, 129
SVC (static VAR compensator), 169–171, 221
SVC stations, 167

T

TCR (thyristor-controlled reactor), 169–171
TCSC (thyristor-controlled series capacitor), 173
Temperature of conductor. *see* Conductor temperature; Conductor thermal modeling
Temperature of transmission line
    effects of elevated (see Elevated temperature effects)
    measured by temperature sensors, 71
Temperature sensors
    on conductors, 18, 19

on transmission lines, 71
Tensile strength loss. *see* Loss of strength
Tension
    and elevated temperature, 74–79, 219
    calculation comparisons, 89–92
    calculation of, 63–66
    in stress/strain relationship, 80
    problem, 75
    real-time rating, 213
Tension monitors on conductors, 18, 21–22
Theory of transmission line ampacity, 27
    differential equation of conductor temperature, 29
    dynamic ampacity, 36–44
    general heat equation, 28–29
    operating conditions, 27
    radial conductor temperature, 47–49
    steady-state ampacity, 29–36
    transient ampacity, 44–47
Thermal conductivity of transmission line conductor, 28
Thermal modeling. *see* Conductor thermal modeling
Thermal ratings of overhead conductors, 67–71
Three-dimensional conductor thermal model, 25
Three-phase powerline magnetic field, 98–100
Thyristor, 164–166, 167
Time series analysis, 25
Time-series stochastic and deterministic models, 74–76
Towers, 6–7
Transient ampacity, 44–47
Transient line rating, 212
Transient stability of generators, 193–195
Transient state operating conditions, 27
Transmission capacity
    enhancement, 12
    factors, 11
    new methods of increasing (see Increasing transmission capacity)
Transmission grid of North America, 3–4
Transmission line
    ampacity theory (see Theory of transmission line ampacity)
    capacity limits (line ratings), 185
    components, 143
    conductor temperature measurement, 71
    electric field (see Electric field of transmission line)
    magnetic field (see Magnetic field of transmission line)
    new technology, 222
    object in computer model, 145–148
    overhead components, 4–8
    rating systems (see Rating transmission lines)
    sag (see Sag)
    security, 74
    tension (see Tension)
Transmission substation, 8–9, 224
Transmission system planning, 195–198
Transmission tap limits, 185–186
TSC (thyristor-switched capacitor), 169–171

U

UHV (ultra-high voltage) lines, 222
Underfrequency relays, 201
Underground transmission, 223
Unsupervised learning in neural network, 129–130
UPFC (Unified Power Flow Controller), 12, **178**, 221

V

Voltage
    and electric field, 108
    calculation of, 104, 205
    listed in computer model, 145
    long-distance transmission, 201
    substation, high-voltage, 8–9
Voltage drop, AC resistance of ACSR conductor, 54–55
Voltage relays, 201

W

Weather conditions
    affect probability distribution of conductor temperature, 74–75
    affecting ampacity, 16, 115
    affecting overhead transmission lines, 67–**68**
    and dynamic line rating system, 188
    and static line rating system, 186–187
    real-time forecasting, 124–126, 213
    required for calculation of equipment rating, 224
Weather dependent line rating systems, 18–19
Weather modeling, 115, 215–217
    Fourier series model, 116–123, 216
    fuzzy set model, 132–137
    neural network model, 127–131, 216
    real-time forecasting, 124–126
    solar radiation model, 137–139
Weather station object in computer model, 148–150
Wind speed, 16
Wind tunnel experiments, 61–63, 212

Index

WindGen, 116, 141, 149, 215
Wire conductors, 4–8

Y

Young's **modulus of elasticity,** 64, 79, 80

Z

Zirconium aluminum alloy conductors, 7–8, **224**